BIOTECHNOLOGY OF FOOD AND FEED ADDITIVES

BIOTECHNOLOGY OF FOOD AND FEED ADDITIVES

Patrícia Alexandra Batista Branco

Delve PUBLISHING

www.delvepublishing.com

Biotechnology of Food and Feed Additives

Patrícia Alexandra Batista Branco

Delve Publishing
2010 Winston Park Drive,
2nd Floor
Oakville, ON L6H 5R7
Canada
www.delvepublishing.com
Tel: 001-289-291-7705
001-905-616-2116
Fax: 001-289-291-7601
Email: orders@arclereducation.com

ISBN: 978-1-77361-204-1 (Hardcover)

ABOUT THE AUTHOR

Patricia Alexandra Batista Branco has received her PhD in Food Engineering from Higher Institute of Agronomy, Lisbon, Portugal and Masters' in Molecular Biology in Health from Egas Moniz School of Health, Portugal. She is an active researcher with many years of experience and has published research papers with reputed Journals.

TABLE OF CONTENTS

LIST OF FIGURES

LIST OF TABLES

PREFACE

This book serves as a reference guide to the biotechnology of food and feed additives and broadly talks about the diverse variety of food and feed additives that are added in human and animal nutrition which addresses the knowledge behind the veil of biotechnology.

In most of the developed nations and developing countries biotechnology is applied both in food and feed industry. Biotechnological methods are applied in production of many food products as preservatives to increase their shelf-life; as sweetening agent to enhance flavor; as a colorant to enhance the appearance of food; and as thickener or emulsifying agent to improve the texture and consistency, as well as the nutritional value of the foodstuff. On the other hand, problems faced by the livestock producers and feed suppliers are even greater. Feed additives are the food supplements that are given to the farm animals that do not get enough nutrients from the regular means and includes vitamins, minerals, amino acids, etc. It is because of the contribution of biotechnology in food and feed industry which makes it possible to enhance the quality and processing of food and feed diets.

The various technological advancements and developments in food and feed industry have made it possible to deliver a better quality and standard of the final product, assuring the safety measures and concerns. The additives in the food we consume and the feed that is added to the animal's diet should be such that it does not lead to any serious health implications. A range of technologies that covers both the conventional and unconventional methods of processing the food and feed additives are applied across the developing nations.

This book covers the potential of biotechnology and its applications in human and animal nutrition to address the safety issues in food and feed systems. In this respect, this book highlights the guidance to the possible and practical usage of additive both in food and feed diets. Hence, overall this book will help and guide us to completely understand the diverse variety of food additive that are used to preserve foods in better conditions, to add aroma and flavor to foods, enzyme and microbial organisms that are benefiting the food, various kinds of thickeners, stabilizers and emulsifiers that are added to enhance the consistency and texture of food, etc., concomitantly feed additives like antibiotics, beta-agonists, prebiotics and probiotics are used to improve the gut environment as well as to match the sufficient nutritional level. At the end, book covers the recent development and advancement in food sectors.

Author

CHAPTER 1

INTRODUCTION TO BIOTECHNOLOGY OF FOOD AND FEED ADDITIVES

"The food you eat can be either the safest and most powerful form of medicine or the slowest form of poison."

— Ann Wigmore

BIOTECHNOLOGY AND ITS ADVANCEMENTS

Biotechnology has romped a miscellaneous title in the field of science and technology crossing many barriers and obstacles. Some of the main achievements in the department of biotechnology are Recombinant DNA technology, Gene transfer, Genetic modifications, Development of hybrid plants, Development of vaccines like- live vaccines, inactivated whole virus vaccines, subunit vaccines, recombinant viral proteins, synthetic peptides, anti-idiotype antibodies and DNA vaccines. Apart from the above attainments, biotechnology also plays a very important and promising role in food processing industry, whereby it assists in attacking the microorganisms present in the bioprocessed foods and hence, upgrading the standards and protection concerning the ingestion of such food products.

Biotechnology has an appreciable application in the food domain be it in refining the texture, edibility and storage of the food; or prohibiting the attack of the food products by various viruses like bacteriophage etc., destroying the unwanted microorganisms by using antimicrobial agents in the food that is toxic to human health, arresting the evolution of mycotoxins; and degradation of other harmful toxins and other toxic elements that are naturally present in the food. This advancement in technology is chiefly concerned with the conventional approach of increasing microorganism's improved strains and further developing the enhanced micro-organic derivatives such as enzymes etc.

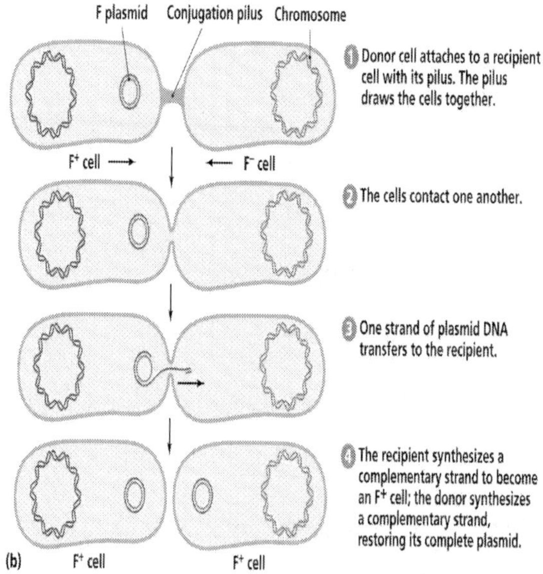

Figure 1.1: Genetic transfer by conjugation

Source: http://biology-pictures.blogspot.in/2011/11/conjugation-in-bacteria.html

Genetic characterization of microorganisms is very important before moving to microorganism's genetic modification to give a clear picture about the favorable and non-favorable factors that are responsible for the growth of the microorganism. Microorganism's improved strains can be produced by numerous processes such as genetic transfer (DNA is transferred from a donor to a recipient) which can be happen in three ways: transformation, transduction and conjugation, , mutagenesis which is the process of generating a genetic mutation on microorganisms by mutagens or genetic recombination. Recombinant DNA technology performs a chief role in microorganism's genetic modification by speeding up the expression of favorable genes and obstructing the expression of non-favorable ones In all the above cases, it is very necessary to understand the genomic study of microorganisms that are associated with the food as it gives a lead in discovering the metabolic process and in genetic mechanisms. From here, genes that are accountable to produce favorable enzymes are identified and proteomics applications are performed for the identification of responsible proteins.

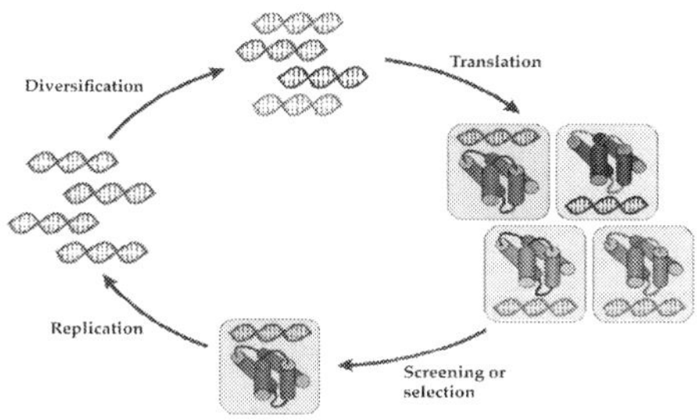

Figure 1.2: The process of diverse genes construction library

Source: http://www.creative-biostructure.com/custom-mutagenesis-services-341.htm

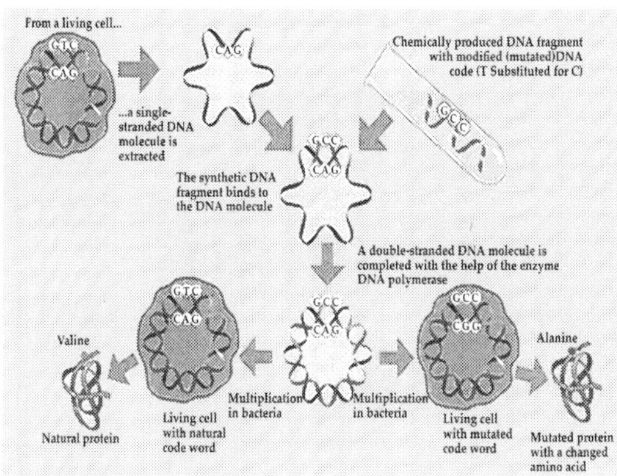

Figure 1.3: Genetic modification by mutagenesis

Source: https://www.nobelprize.org/nobel_prizes/chemistry/laureates/1993/illpres/site.gif

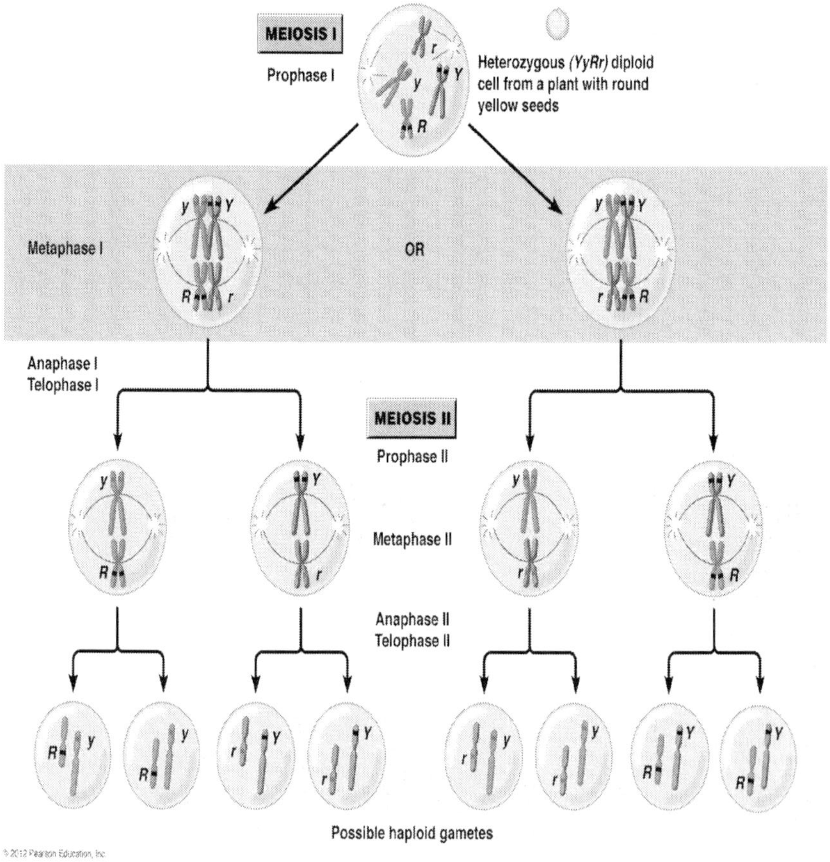

Figure 1.4: Genetic recombination by hybridization

Source: http://www.mun.ca/biology/desmid/brian/BIOL2060/BIOL2060-20/CB20.html

Biotechnology is a very broad field in terms of its processing and applications; it also participates and engage a key role in protein engineering. Protein engineering is the process of commercial production of beneficial enzymes that are present in the microorganisms at a large scale by microbial culturing of microorganisms in tanks. The bacteria that are multiplied by industrial culturing methods and enzymes that are assembled as metabolites are isolated and can be used in food industry.

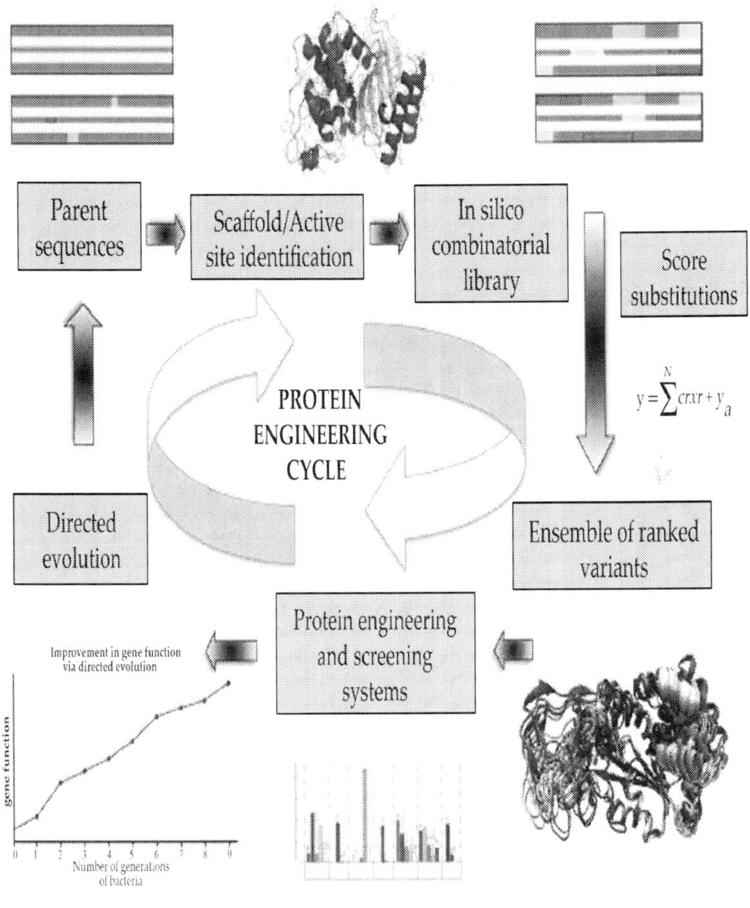

Figure 1.5: The protein engineering cycle and design

Source: https://www.slideshare.net/pablocarb/lecture01-6180459

Since the enzymes are produced massively, in large tank reactors, this makes them easily accessible for usage at a low cost and the quality is maintained. Moreover, with the help of genetic modification of microorganisms it has become possible to make modified enzymes with improved efficacy. Enzymes improved by genetic modification techniques are superior to usual enzymes in terms of their thermal stability and optimal pH. .Biotechnology assists an easy access to understand the metabolic pathways of several different microorganisms and this in turn helps in isolating the essential derivatives.

Furthermore, the technology also supports the identification of different pesticides, pathogens, anti-nutritional factors and several other contaminants like mycotoxins that otherwise exists in the food and causes toxicity and results in diminishing the shelf life of food, these tests are performed by various techniques like ELISA (enzyme-linked immune sorbent assay), microarray (commonly known as DNA chip or biochip), etc.

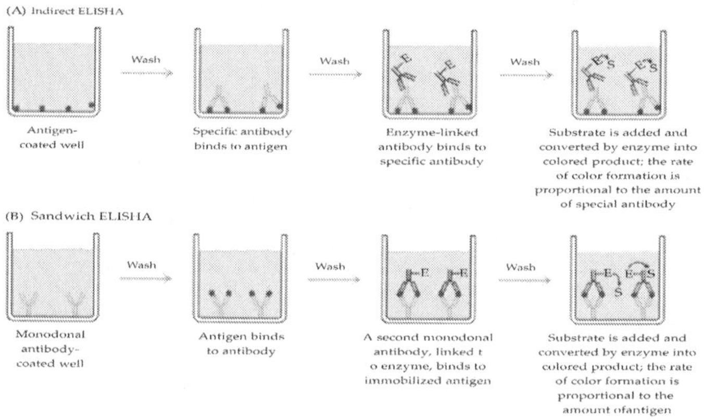

Figure 1.6: Steps of indirect and sandwich ELISA

Source: http://www.namrata.co/elisa-a-quick-review/steps-of-elisa/

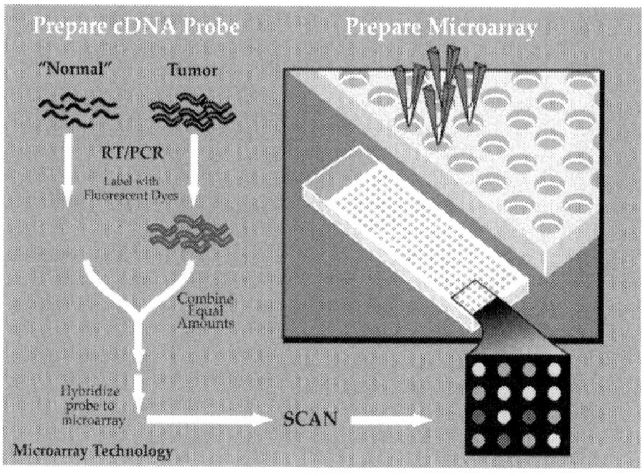

Figure 1.7: DNA microarray chip

Source: https://www.genome.gov/10000533/dna-microarray-technology/

In October 1985, at an international symposium the interface link between biotechnology and food processing was explored. Although, biotechnology has a very significant role in food processing industry still its progress is very limited to developing countries in the past few years due tosocio-economic factors. Food processing industry and research in this field can only grow when the government raises funds in this sector and support the development strategy of food industry. Food processing is the method of converting the simply biodegradable and non-palatable raw materials into edible foods, with longer shelf life. Several amino-acids, food additives, food-flavoring agents and preservatives are the diverse by-products obtained from varied microorganisms. It is due to the numerous advancements in biotechnology in the past several years that made it very facile to understand the metabolic pathways of various microorganisms and consequently helps the isolation of the needful derivatives.

Food industry is a combination of complex and diverse operations that are responsible for processing the food in such a way that it gives a palatable taste as well as addition of certain preservatives that enhances flavor and increases the shelf-life of food. Food processing utilizes several unit actions and technologies to transform the bulky, biodegradable and unpalatable raw materials into edible foods with increased shelf-life so they can be used for longer time. By the term, "processing" it means to secure the food by diminishing the waste and losses that are present in the food chain, and to increase the food marketability and availability. Besides, food is also processed to enhance its taste and quality, as well as food safety.

Food safety is a scientific regulation that describes the handling, storage and preparation of food. Besides, guarantees that the food served in the market would not lead to any foodborne illness to the consumer, when it is consumed in intended amount. There are number of practices that should be followed to circumvent possible health hazards. The safety of food should flow between the industry and the market and from market to the consumer. Regarding the flow of food from industry to the market, there should always be some directions which need to be followed for maintaining the quality and safety of food. Some of the practices that comes under this are food hygiene, labelling of food, food additives and pesticide residues, several strategies on food and biotechnology, as well as government inputs and guidelines on import and export assessment and record systems for foods.

With the increase of these economic sector in many countries, consumers now prefer processed foods more than the primary food. This leads to immense growth of food processing and packaging business and the numbers have reached several trillion dollars over the past years. Approximately 16 million people work in this industry as per the report. Today, food processing is not just about assembling productivity requirements but also to acquire enlightened automation, monitoring and control procedure and discipline. Some of the research and developments related to significant total of food processing that demands automation study, monitoring and control are:

- Improvement in food safety and reliability
- Micro and macro structure of food
- Bio-encapsulation and microencapsulation of food
- GMOs (genetically modified organisms)
- Development in chemistry and scientific research.
- Betterment in present techniques.
- Mechanisms used for oral destruction of food
- Novel foods and supplements
- Development of new product in the market
- Food innovation
- Nutraceuticals

GMPs	Sanitation	Regulatory Compliance	Quality Control	HACCP
KPIs	**KPIs**	**KPIs**	**KPIs**	**KPIs**
Training	Training	Training	Training	Training
Documentation	Documented Procedures	Documentation	Control of Documents	Documentation
Specifications	Cleaning Schedules	Plant Registration	Specifications	Product descriptions
Process Controls	Monitoring	Process Filling	Testing Methods	Flow Charts
Equipment	Actions on Deviations	Process Authority	Documented Procedures	Hazard Analysis
Manufacturing Procedures	Chemical Control	Inspection Readiness	Calibration and Testing	Critical Control Points
Components Handleing	Cleaning Equipement	Recall Procedures	Inspection and Testing	Critical Limits
Traceability	Water Supply	Recordkeeping	Process Controls	Monitoring
Lot Coding	Sewars	Auditing	Actions on Deviations	Actions in Deviations
Storage Facil;ities	Process Waste		Recordkeeping	Responsible Personnel
Buildinga and Grounds	Resfuse Handling		Auditing	Verification
Maintanance Procedures	Recordkeeping			Recordkeeping
Welfare Facilities	Auditing			Auditing
Pest Control				
Chemical Control				
Vendor Programes				
Co-packer Programs				
Recordkeeping				
Auditing				

Figure 1.8: Key performance indicators for rating elements of a food safety system

Source: http://www.foodsafetymagazine.com/magazine-archive1/december-2002january-2003/the-total-plant-food-safety-audit-rating-your-overall-system/

Biotechnology in food and feed industry

Application of biotechnology in food processing industry aims at selection and manipulation of microorganisms with the target of boosting the process control, product consistency, quality, safety, and yield while elevating process efficacy.

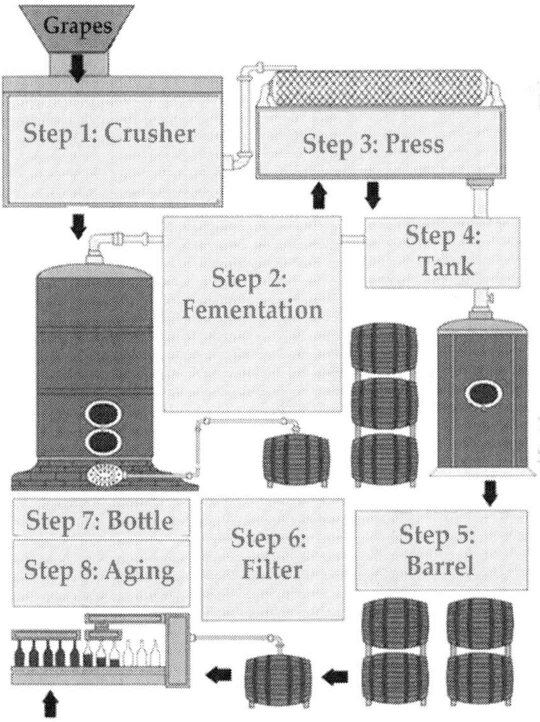

Figure 1.9: Fermentation and processing of wine

Source: https://www.softchalk.com/lessonchallenge09/lesson/wineTutorial/wineTutorial4.html

Biotechnology in food processing industry is used for the preservation of food with the help of varied type of microorganisms, this sector also assists the production of wide range of value-added products like flavored compounds, microbial cultures, enzymes, vitamins and other food ingredients. For the improvement of microbial cultures used in food processing applications some of the traditional methods of genetic improvement are applied such as classical mutagenesis and bacterial conjugation. Both bacterial conjugation and mutagenesis main focus is on upgrading the quality and yield of microorganisms and metabolites. The yeast cultures required in brewing, baking, and in beverage production are improved applying genetic recombination by hybridization. The strains of Saccharomyces cerevisiae have been mainly studied for enhanced the wine fermentation process, bio-preservation and processing abilities, as well as to increase the quality (sensory quality) of wine.

The foremost term introduced in modern biotechnology i.e. recombinant gene technology is chiefly used for strain improvement. Although the accessibility of genetic manipulation tools that intended to improve the microbial cultures linked with food fermentations are fortified by regulatory concerns and consumer perceptions. The genetically modified microbial cultures are still utilized to produce various enzymes and food processing ingredients like polyunsaturated fatty acids, monosodium glutamate, and some amino acids. Biotechnology is broadly engaged as a diagnostic tool instructing to supervise the food safety, preventing and diagnosing the foodborne illness and verifying the emergence of foods.

The increased cost and lack of animal feed and feed ingredients lead to the need to advance and expand the feed utilization. With the on-growing biotechnological advancements in the past years, the use of non-toxic fungi to expand the fibrous feeds has been allowed. Biotechnology is still in the early stages of development in animal feed production. To change the rumen fermentation, an extensive variety of feed additives are used other than the antibiotics. Some of the feed additives that are needed for rumen fermentation includes- enhanced propionic acid production, reduced protein degradation, basic components that are needed for methanogenesis reduction, improved microbial protein synthesis as well as protozoan inhibition. Antibiotics, microbes, and specific oligonucleotide substrates are these amongst such substrates. Further, the enzymes are prepared and produced effectively in large quantities, that too inexpensively. Due to the lack of better feed quality and management system in developing countries, the developed countries have higher probability to produce better animal nutrition. "The goal of using biotechnology in animal nutrition is to improve the plane of nutrition through increasing availability of nutrients from feed and to reduce the wastage of the feed."

APPLICATIONS OF MICROORGANISMS IN BIOTECHNOLOGY

Generally, microbial inoculants are used when one talks about food processing and food biotechnology in developing countries to amplify the properties like texture, fragrance, taste, shelf-life and nutritional evaluation of foods. The process by which the microorganisms and its enzymes bring about the beneficial changes in food medium is called as fermentation. The fermentation process is broadly helpful for microbial culture production, enzyme production, flavors, food additives and chain of elevated value-added products. These value-added products are basically produced in countries that are technologically very advanced and strong in terms of economic values and sources, these products are used in both food and non-food processing applications. The developing countries import these high value-added products for food processing applications from the developed countries where they are produced.

Microorganisms are used in food industry since a very long time, much before the fermentation process being known, microorganisms are used for preparing variety of food products and beverages. Food products like bread, cheese, yoghurt, curd etc. are all microbial preparation. Other than that alcohol beverages like beer and wine also results of microbial preparation. Some microbial species are helpful in expansion or development of peculiar flavor which is unique to certain specific wines. Louis Pasteur was the first to understand and explain the importance of microorganism in food based and industrial biotechnology. He has shown the role of microorganisms in spoilage and illuminate that microorganisms are also involved in fermentation.

There are several amino acids like lysine and glutamic acid that are used in the food processing industry as a nutritional supplement to elevate the nutritional value and also to amplify or enhance the compound by adding flavor to it. Mutants of Corynebacterium glutamicum are used in large quantities to produce glutamic acid, this bacterium, responsible for glutamic acid production, is insufficient to transfigure the intermediate of tricarboxylic acid (TCA) cycle alpha ketoglutarate to succinyl-CoA. Production of lysine is also from mutants of Corynebacterium glutamicum which obstructs the homoserine combination and accumulates lysine.

Other than the amino acids that are used in food industry organic acids like citric acid, acetic acid, lactic acid, fumaric acid and gluconic acids are also the vital commodities of food industry. Citric acid has a broad range of applications in food and pharmaceutical industry. It is the most essential organic acid that is produced by fermentation. for citric acid production Aspergillus niger is the microorganism most used. Other organic acids like gluconic acid, acetic acid, lactic acid are also the principal organic acids of food industry produced by microorganisms.

Lactic acid is another significant acid that is produced by lactic acid bacteria and it can also be synthesized chemically in equal amount. Lactic acid is chiefly utilized in manufacturing emulsifiers and as food additives in food processing industry. Lactic acid occurs in two forms i.e. D (-)- and L(+)- form, these are the two enantiomers of lactic acid. The chemically synthesized lactic acid produces DL-lactic acid, whereas L-lactic acid is produced by fermentation and is mainly required for normal human metabolism. For lactic acid production stains of bacterium like Lactobacillus delbrueckii, L. casei, L. acidophilus etc. are used for commercial fermentation.

Another important organic acid i.e. gluconic acid is produced by the fungus Aspergillus niger by the bacteria Acetobacter suboxydans and also by some species of Penicillium.

Table 1.1: Biotechnology products for the food industry

Product	**Utilization**
Enzymes	
Amylase	High fructose corn syrup
Isomerase	Production of high fructose corn syrup
Rennet	Cheesemaking
Proteases	Meat tenderizer
Pullulanase	Light or "Lite" beer
Organic acids	

Citric acid	Acidulant
Benzoic, propionic acid	Food preservative
Amino acids	
Methionine, lysine, tryptophan	Nutritional supplement
Aspartic acid, phenylalanine	Aspartame production
Product	**Utilization**
Vitamins	Nutritional supplement
Low calorie products	
Aspartame, thaumatin, Monellin	Non-nutritive sweeteners
Modified fatty acids and triglycerides	Food additives and cooking oil
Microbial polysaccharides	Stabilizers, thickeners, gelling agents
Flavors and pigments	Flavoring and coloring agents
Single cell protein	Animal and human food supplement

Microbial enzymes in food industry

Microbial enzymes are used in food industry in fermented foods such as bread and cheese. They are also used in fruit juices, milk and clotting, etc. food processing industry mostly uses purified enzymes. However, to produce alcoholic beverages living cells (yeast) are necessary to perform alcoholic fermentation. According to the report on global market trends, the world market of microbial enzymes is over 1.5 billion dollars. Enzymes are also gaining importance in several other fields like in paper and pulp industry, textile industry, chemical industry, detergents etc. Nevertheless, the largest application of enzymes is seen in food processing industry till date.

Use of enzymes in the food industry

Process	Use
Baking	Enzymes in yeast convert **sugar** → ethanol and CO_2 CO2 makes the bread dough **rise**
Brewing	Enzymes in yeast convert **sugar** → ethanol and CO_2 Ethanol makes the drink alcoholic CO_2 makes the drink fizzy
Chees making	Enzyme **rennin,** extracted from cow's stomachs, is used to clot milk. Rennin can now be made using genetically engineered bacteria.
Making baby foods	Trypsin (a protease) is used to predigest baby foods.

Figure 1.10: Use of enzymes in food industry

Source: http://biology-igcse.weebly.com/use-in-the-food-industry.html

Production of cheese:

Traditionally, cheese was being produced by using calf rennet, which is a complex of enzymes. Then, gradually with time, there was severe shortage seen in calf rennet, to overcome this loss several substitutes omicroorganisms that produce rennin (enzyme that is also present in calf rennet such as Rhizomucor miehei, Endothia parasitica and Rhizomucor pusillus were selected by mutation and selection method.

Nowadays, several microorganisms such as Escherichia coli, Aspergillus niger and Kluyveromyces lactis are genetically modified (GMOs, genetically modified organisms) in order to obtain rennin. These GMOs contain the gene encoding chymosin also called rennin (enzyme present in calf rennet).. The gene encoding chymosin is extract from the stomach cells of ruminant mammals (calves), and inserted into a plasmid, and then inserted into microbial cells. As a result, the microorganism starts to produce chymosin or rennin, the key component of calf rennet. Cheese comes in numerous variabilities. The range defines the ingredients, characteristics, and processing of the cheese. Cheese is being prepared by using pasteurized or raw milk. Cheese prepared from raw milk conveys diverse flavors and texture appearances to the finished cheese. In few cheese ranges, before pasteurization, raw milk goes under a mild heat treatment former to cheese making so as to destroy some of the spoilage microorganisms and deliver better environments for the cheese cultures. Cheese that is made from raw milk should be aged for at least 60-65 days, as defined in the Code of Federal Regulations (CFR) guidelines, so as to lessen the likelihood of contact with disease-causing (pathogenic) microorganisms that may exist in the milk. However, there are also few varieties that must be aged longer than 60-65 days.

Cheese is broadly classified into two types:

a) Acid and rennet cheese,

b) Natural or processed cheese.

Acid cheese is prepared by adding acid to the milk to cause the proteins to coagulate. Fresh cheeses, such as cream cheese, are prepared by direct acidification. Many types of cheese, like swiss or cheddar, they use rennet enzyme besides the starter culture to coagulate the milk. The term "natural cheese" refers to the cheese that is made directly from milk. Processed cheese is prepared by using natural cheese and other ingredients that are cooked together to alter the textural or melting properties and upgrade the shelf life.

Figure 1.11: Distinct types of cheese

Source: https://www.google.co.in/search?q=natural+cheese&source=lnms&tbm=isch&sa=X&ved=0ahUKEwjUid2ekpzVAhVELY8KHd2VDLwQ_AUICigB&biw=1366&bih=662#tbm=isch&q=different+types+of+cheese&imgrc=W9LQaGwajk5o7M:

Ingredients - The key ingredient in cheese is milk. Cheese is being prepared by using cow, goat, sheep, water buffalo or a mixture of these milks.

The type of coagulant used is directly dependent on the type of cheese desired.

Example- for acid cheese, source such as- acetic acid, an acid in vinegar, or Glucono-delta-lactone, a mild food acid is utilized.

For rennet cheese, calf rennet or a rennet that is prepared through microbial bioprocessing is utilized. Sometimes, calcium chloride is added to the cheese to boost the coagulation properties of the milk.

Flavors are added some of the times depending on the type of cheese. Some of the most common ingredients that are used in flavorings include herbs, spices, hot and sweet peppers, horseradish, and port wine.

Bacterial Cultures - Cheese making cultures are called lactic acid bacteria (LAB) as their prime source of energy is the lactose in milk and their principal metabolic creation is lactic acid. There are extensive bacterial cultures available that deliver diverse flavors and textural features to cheese. In the cheese making process, starter cultures are used early so as to assist coagulation by lowering the pH preceding to rennet addition. The metabolism of the starter cultures helps in giving the needed flavored compounds, and also support to avert the growth of spoilage microorganisms and pathogens. Some of the most common starter bacteria are-

Lactococcus lactis (sub-species lactis or creminis), Lactobacillus helveticus, Lactobacillus delbrueckii (sub-species bulgaricus), and Streptococcus salivarius (sub-species thermophilus). Adjunct cultures are used to deliver or improve the distinctive flavors and textures of cheese. Some of the most widely used adjunct cultures added during manufacturing of cheese includes- for giving flavor to cheddar cheese, lactobacillus plantarun or lactobacillus casei are used, for eye formation in swiss cheese, Propionibacterium freudenreichii are used. Adjunct cultures are also used as a smear, that are used for washing the outer part of the formed cheese, like- use of Brevibacterium linens of brick, gruyere, and limburger cheese. Molds and yeasts are also used in cheese such that they provide a characteristic color and flavors to diverse varieties of cheese. For example, Torula yeast is used in the smear for the maturing of limburger and brick cheese. Several examples of molds are Penicillium camemberti in camembert and brie, and Penicillium roqueforti in blue cheeses. Below is the overall Cheese processing steps-

- Standardization of Milk- Milk is frequently standardized before cheese making begins to adjust the protein to fat ratio to make a decent quality cheese with a high yield.
- Pasteurization of Milk- The milk is pasteurized or mildly heat-treated, depending upon the desired cheese, to decrease the quantity of spoilage organisms and recover the environment for the growth of starter cultures. Some of the milk varieties are made from raw milk, and hence they are not pasteurized. Raw milk cheese should at least be aged for 60 days to lessen the possibility of getting in contact with any disease-causing microorganisms that are present in the milk.
- Cool Milk- After pasteurization, milk is cooled to in order to

reach the temperature required for the starter bacteria to grow. . In case of raw milk, temperature is 90°F (32°C).

- Inoculation with Starter & Non-Starter Bacteria- Both starter and non-starter bacteria (previously isolated from milk) are added to the milk and held at 32°C for 30 minutes to ripen. The ripening permits the bacteria to grow and start fermentation, which in turn drops the pH and advances the flavor of the cheese.
- Addition of Rennet and Formation of Curd- The rennet is the enzyme which acts on the milk proteins and assists in formation of the curd. After the addition of rennet, the curd is kept undisturbed for approximately 30 minutes such that a firm coagulum is formed.
- Cut Curd and Heat- The curd is permitted to ferment till it reaches a pH 6.4 once the pH of the cured is 6.4, the curd is cut with cheese blades into small pieces and heated to a temperature of 38°C. The heating assists to disperse the whey from the curd*.
- Drain whey- The whey is then drained and the curd forms a mat.
- Texture of curd- The curd mats are cut down into segments and loaded on top of each other and flipped periodically. This complete process is known as cheddaring. With the help of cheddaring it is possible eject more whey, permits the fermentation process to continue until a pH of 5.1 to 5.5 is stretched, and lets the mats to knit together to form a tighter matted assembly.
- Dry Salt or Salt water solution- For the preparation of cheddar cheese, milled curd pieces are set back in the vat and sprinkling is done with dry salt on the curd and mixing in the salt. In some of the cheese forms, like mozzarella, the curd is shaped into loaves and then the loaves are positioned in salt water solution.
- Formation of Cheese into Blocks- The salted curd pieces are sited in cheese hoops and compelled into blocks to form the cheese blocks.

- Store and Age- after the formation of blocks, the cheese is then stored into coolers till their desired age i.e. from several months to several years, depending upon the variety of cheese, is reached.
- Packaging- Cheese is then cut and packaged into blocks, and is ready to be sold in the market.

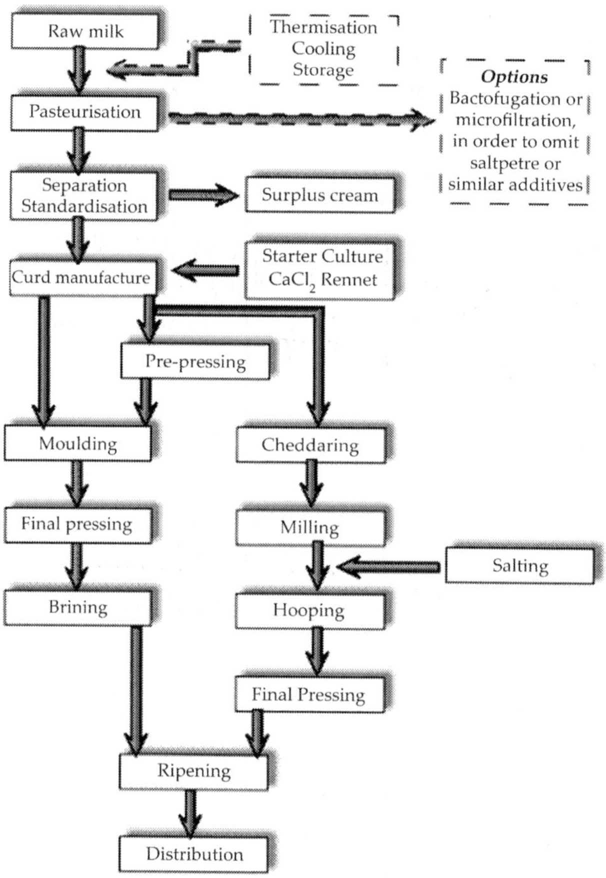

Figure 1.12: Production of Cheese

Source: http://www.scq.ubc.ca/food-microbiology-the-basics-and-the-details-of-cheese-production/

- Fermentative production of amino acids: Food industry uses many forms of amino acids for enhancing the flavor, as a sweetening agent, and as food and feed additives. Some amino acids like L-glutamate is used as a flavor enhancer, glycine is used for adding sweetness to foods, aspartate and phenylalanine is used as low-calorie sweetener, and lysine and methionine as food and feed additives, etc. Both chemical and enzymatic amino acids have a total commercial production of 1.6 million tonnes, out of which glutamate serves 1 million tonnes, after which lysine and methionine and each serve around 350,000 tonnes. The market is continuously growing at the rate of 5-10% every year. As glutamate is produced in larger quantities, it is the cheapest amino acid followed by lysine and methionine.

Corynebacterium glutamicin is chiefly used for the commercial preparation of glutamate, phenylalanine, threonine, lysine, etc. Other microorganisms like Escherichia coli, Serratia, Bacillus, Candida sand Saccharomyces. are also used for preparation of amino acid, some of which are also genetically modified organisms. Due to the regulatory control over amino acids synthesis by the microorganisms a large amounts of amino acids is not accumulated

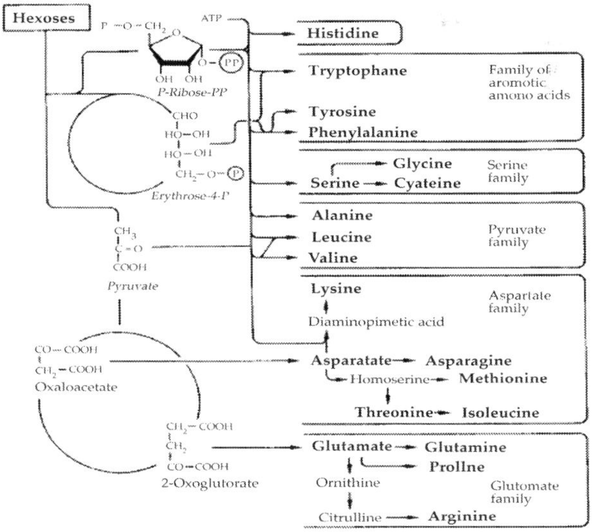

Figure 1.13: Amino acid production pathway

Source: https://www.slideshare.net/guest8bfdff/lysine-production

Biotechnology of dairy products:

Bacteria's like Lactobacillus, Lactococcus, Bifidobacterium, Leuconostoc, and Pediococcus are used in dairy products. They are used to enhance the flavor, improve the texture, preservation and nutritional value of dairy products as well as of cereals, vegetables, and legume fermentation products which includes cheese, buttermilk, yoghurt, etc. Some of the bacteria are even used as probiotics, contributing to improve the overall health of the consumer. Other lactic acid bacteria like Lactobacillus acidophilus and Lactobacillus sake, produce antimicrobial bacteriocins, which helps in managing the growth of unwanted microorganisms. Genetically engineered lactic acid bacteria are also prepared by carefully studying the molecular strategies. These genetically engineered lactic acid bacteria have improved fermentation efficacy, better shelf-life, sensory and nutritional properties for the product, etc.

The undesirable and unwanted impurities present in the cheese and yoghurt when they were made leads to degraded flavor, less yield and also leads to food poisoning. The genetically modified lactic acid bacteria grow faster than the impurities or contaminants present in dairy products. These genetically engineered bacteria inhibiting the growth of certain bacteria that causes contamination by production of antimicrobial agents.

Milk is a respected nutritious food that don't have a longer shelf-life and needs watchful handling. Milk is extremely delicate because it is a optimal medium for the growth of microorganisms, that leads to spoilage and diseases in consumers.

Milk processing consents the safeguarding of milk for days, weeks or months and helps to diminish food-borne illness. The functioning life of milk can be prolonged for several days through techniques like- cooling. By the help of pasteurization, the usable life of milk is extended and number of pathogenic bacteria's that are responsible for causing health hazards are decreased up to certain extent.

Sometimes, milk is processed further so that it can be converted into more concentrated, effortlessly transportable and high value dairy product with increased shelf life, for example- ghee, butter and cheese. Dairy product processing stretches the income of small-scale dairy producers more than selling raw milk and proposes better chances to grasp regional and urban markets. Milk processing also helps to agreement with seasonal fluctuations in milk supply. The conversion of raw milk into processed milk and products can profit whole community by making off-farm jobs in milk collection, transportation, processing and marketing.

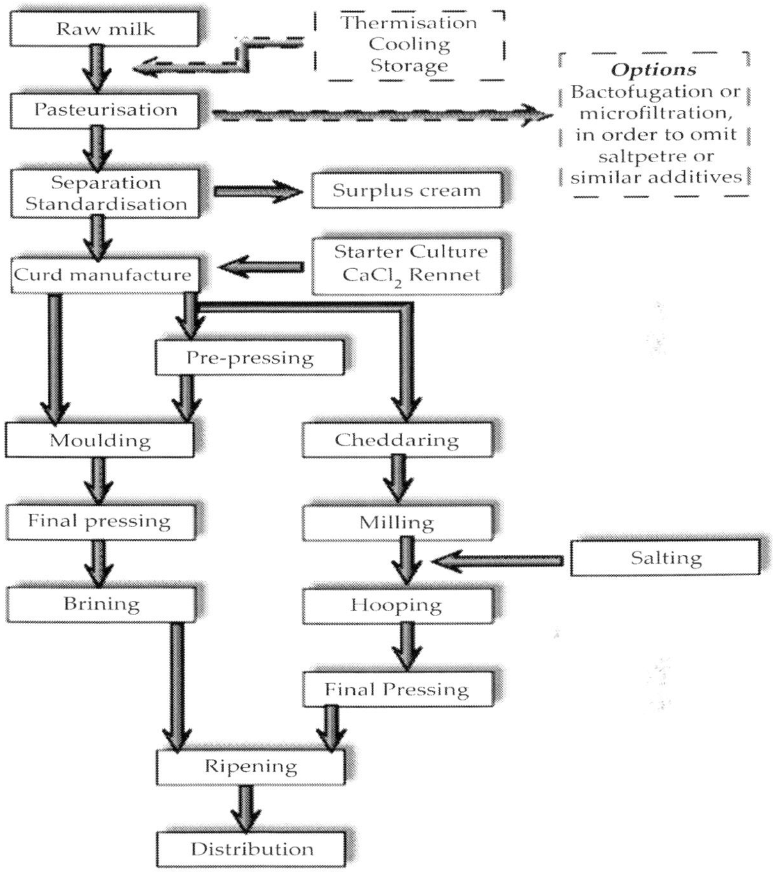

Figure 1.14: Production of Milk

Source: https://cbecl.files.wordpress.com/2012/07/dairy-milk-processing-flow-chart.jpeg

Alcoholic fermentation using improved cultures:

Several stains of yeast that are used in beer brewing are picked based on the aroma and flavor they impart during fermentation. Factors like osmotolerance, tolerance to low pH, fermentation rate, high ethanol tolerance are other important characteristics that contemplate several different stains. Saccharomyces cerevisiae and Saccharomyces uvarum are mainly used for commercial production of beer. However, brewing beer don't use genetically modified organisms for commercial purpose, but presence of genetically modified organisms helps to understand the genes that controls the brewer's yeast properties and productivity. Microbial fermentation is the main source for the current production of food formulation, some of the products, like enzymes, amino acids, organic acids are the result of microbial fermentation. The future of biotechnology is very bright as it is growing each passing day and is the most organized sector. In the coming years, advanced biotechnological techniques will be used to design the microorganisms that are efficient to produce high-value additives both in terms of capability and cost.

For alcoholic fermentation, microorganisms such as- Saccharomyces cerevisiae and Saccharomyces uvarum are mainly involved as discussed in the previous paragraph. The main stages of beer- making involves:

Malting

As soon as the barley is grown and harvested, malting process begins. Malting is done such that the enzymes in the grain are formed which are needed for the conversion of starch into sugars. This is how sugar is obtained from the grain in the brewing process. The process of mating begins by soaking the grains in water, when these grains reaches the proper level of germination, then they are dried with hot air and the malting process is then stopped. This is how the malts like- crystal, chocolate, and amber are obtained.

Figure 1.15: The barley seeds

Source: http://www.hcexport.in/barley_seeds.html

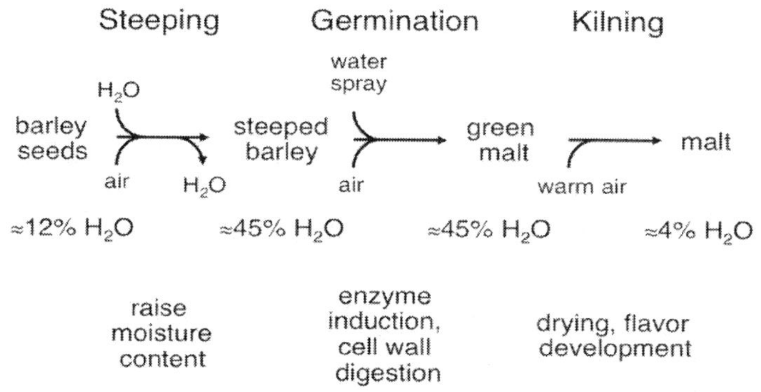

Figure 1.16: The malting process

Source: https://justbeerapp.com/article/beer-styles-the-ingredients-part-one-malt

Milling

Once the malted grains are dried, they are stored in a sealed container for about one year in un-milled form. Milling is a process in which the outer layer of the grain is crushed, and the inner layer is broken letting the mashing liquor to better access the entireness of the grain.

There are various sizes of mills, like- smaller mills, that are hooked up with the power drill, and are basically used by home brewers and brew shops.

Figure 1.17: The barley crusher

Source: http://craftbeeracademy.com/malted-barley-and-beer/

Figure 1.18: well-milled barley

Source: http://craftbeertemple.com/videoblog/brewing-process/

Mashing

The freshly milled barley goes further for mashing process. For malting process to begin, the milled grains are added to large vessels called the mash tun, further they are mixed with warm water to form the mash. The heat generated activates the enzymes inside the barley. The enzyme then converts the starch in the grain into sugars.

Since there are different types of enzymes that are present in the barley, each with an ideal temperature, the mash temperature monitored is extremely close. Brewers can control the types of sugars produced by enzymes by raising and lowering the temperature of the mash. When the temperature is low, highly fermentable sugars are formed which results in dry beers. Whereas when the temperature is high, it takes enough time for the sugars to be digested by the yeast, this results in beer with some unfermented sugars left behind, which results in sweeter, full-bodied end-product.

The enzymes that participates in the mashing process are very fast, within an hour they successfully convert starch to sugar, as soon this point is reached the enzymatic activity of the enzymes are stopped by raising the temperature of the mash to 200 degrees Fahrenheit, a process called mashing out.

Figure 1.19: The mashing process

Source: https://beerconnoisseur.com/articles/beer-101-fundamental-steps-brewing

Lautering

The next step in beer production is lautering, in this process the resulting mash is taken and is ready to be separate out the spent grain from the sugary liquid known as wort. as soon as the processing of lautering begins, the mash formed is shifted to the vessel with a false-bottom known as lauter tun. In this step, the cleared wort is drained away from the barley grist and hulls. In order to extract more fermentable sugars from the grain, water is continuously added, this is called as sparging. Sparging must be done slowly so that it does not disrupt the grain bed which is a natural filter for the wort. Generally, the sparge water is added at the same rate as the wort is being drained beneath while some of the brewing methods includes complete drainage of wort, then water is added doing it all over again. This round of lautering is known as second runnings, which was traditionally made for small beers.

Figure 1.20: The above image depicts the lautering process

Source: https://beerconnoisseur.com/articles/beer-101-fundamental-steps-brewing?page=1

Boil

As soon as the wort is separated out from the grains, a strong and prolonged boiling begins for about one to two hours. The boiling process is for various reasons. Firstly, from the traditional point of view, it sterilizes the beer. Even today, it is considered as a very important process, but in historical times, this process was like a life-saver. It is due to this reason, in early times people use to drink beer including small children because they believed that something is there in the brewing process which makes beer healthy. However, boiling does not only help in sterilizing the brew, but there is lot more involved. The hops introduced in the boiling water breaks down or isomerizes, which alters the composition of the acids within the hops and hence releases bitterness to the beer. In order to lend more bitterness to the brew, the hops are boiled for longer time, and due to this more alpha acid are isomerized, as a result more bitterness. Hops that are added to lend more bitterness to the brew are known as early hop additions, and are usually acceptable to boil in the beer for an hour or so. Later, the hops that are added don't add much bitterness, instead they lend aroma and flavor to the beer. The reason behind this is that in addition to alpha acids, hops also comprises of high volatility and pungent oils, that breaks down very quickly in water. Nonetheless, if hops are exposed to the boiling wort at low temperatures and for a shorter time duration, they will release these oils plus give aromatic flavor to the beer. Brewers usually add hops about 15-20 minutes before the end of the boil, to extract flavor. To get the aromatic smell, hops are even added later, just 2-3 minutes before the boiling ends. Sometimes, aromatic hops are added after the end of the boil. As the boil is finished it is directed to a whirlpool to assemble and collect the hop matter and the accumulated coagulated proteins. It is then immediately cooled. Cooling is a very important step, as the beer begins to oxidize and produce off flavors at high temperatures.

Figure 1.21: Hops that are added during boiling process at various stages

Source: https://beerconnoisseur.com/articles/beer-101-fundamental-steps-brewing?page=1

Fermentation

A soon as the wort is cooled, it is sent to the fermenter, for fermentation. Fermenter is nothing but a large stainless-steel vat. Once the cooled wort is added to the fermenter, next step is addition of yeast, the moment yeast is added to the fermenter it starts to degrade the sugars that were created during the mash process. The degradation of sugar from the wort by yeasts results in carbon dioxide and alcohol (alcoholic fermentation), and a variety of flavoring compounds that vary critically depending on few variables namelyspecific strain of yeast, and fermentation temperature of beer. The fermentation time varies significantly depending on the type of yeast used, fermentation can differ from few days for simple ale, to months for lagers. The brewing of the beer varies considerably depending on path the brewer decides to go. If brewer desires to brew an ale, then the brewer will pitch for ale yeast, with the preferable fermenting temperature of 65-75 degrees Fahrenheit, which is the ideal temperature for ale yeast. Ale yeast is also known as top-fermenting yeast and it forms a thick layered pillowy foam at the bottom of the vessel.

If brewer wants to brew a lager, the brewer needs to keep the beer much cooler, typically between 45-55 degrees Fahrenheit. The lager yeast works slowly and efficiently at these temperatures in order to produce a clean tasting beer with little flavor that is given by the yeast itself. The lager yeast or bottom fermenting yeast, don't form a pellicle at the top of the vessel, like ale yeast.

Figure 1.22: Ale yeast forming a high krausen in an open fermenter

Source: http://craftbeertemple.com/videoblog/brewing-process/

Conditioning

As the yeast consumes all the sugars that are available, the first step of fermentation i.e. primary fermentation is over. Technically, the beer is no longer fermenting, but the yeast work is continued. In the middle of fermentation, some off-flavor compounds are created with the production of alcohol and carbon dioxide. Some of the most common off-favored compounds such as sulfur, butter and green apples, are associated with under-conditioned beer. For ales, conditioning process is usually done in a week or so, whereas in lagers, where the yeast works bit slowly, it may take a month or so for this process to be completed. To speed up the process of conditioning in lagers, brewers employ a practice named as kräusening, in this process the fermenting wort is still added to condition the beer and help to kick start the conditioning process. As the fermentation is completed, and the yeast have operated their way through the off-flavors, it becomes inactive and deposits in the bottom of the fermentation vessel. The brewer can simply remove the yeast, from here, leaving only clear, bright beer. Some brewers filter the beer mechanically, or by means of filtering agents, which are added and assists in collecting yeast into larger groups, which then results in reduction in suspension at a faster rate.

Figure 1.23: Conditioning tanks for conditioning of beer

Source: http://alliedbeveragetanks.com/wp/brewery-tanks/

Packaging

The final step in the brewing process is packaging the beer. Packaging is simply the process of transfer the finished product into vessels for mass consumption. Mostly, the beers are packed into cans, bottles, and kegs. Many of the brewers also force carbonates into the beers just before the packing, so that any carbon dioxide developed during fermentation can be escaped out into the atmosphere. If the carbon dioxide is not escaped out, it leads to potential rupturing of the tanks due to the pressure that is build-up. Some brewers focus on a different concept, named as bottle conditioning, to avoid the rupturing of the tanks due to the build-up pressure, for this they add yeast and sugar to the beer after the beer has been packaged, this is done to bring the second fermentation inside the bottle, thus adding depth of flavor, and carbonation. Bottle conditioning also improves the shelf-life of beer, as the yeast consumes all the oxygen that is oozed into the bottled beer.

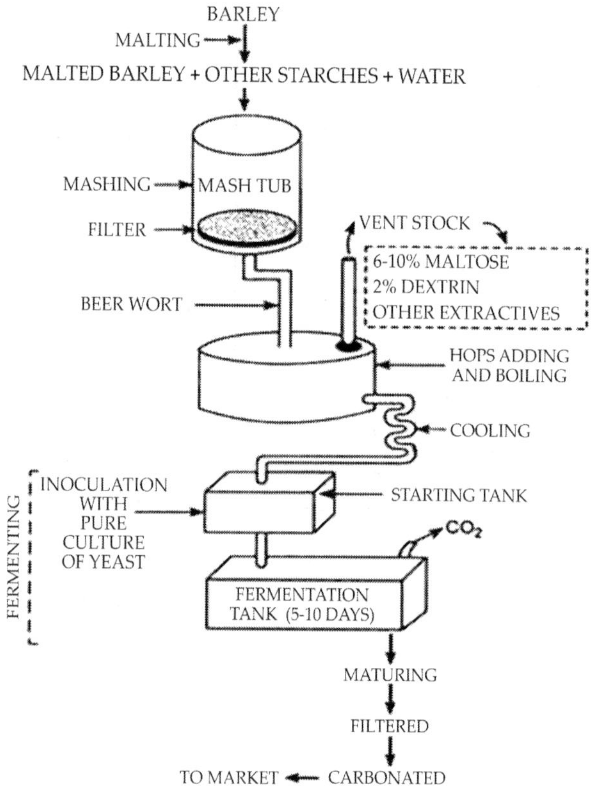

Figure 1.24: commercial production of beer

Source: http://www.biologydiscussion.com/industrial-microbiology-2/alcoholic-beverages-undistilled-and-

REFERENCES

1. Biotechnology of Food and Feed Additives. (2014). Advances in Biochemical Engineering/Biotechnology.
2. Biotechnologyforums.com. Biotechnology in Food Processing Industry. [online] Available at: http://www.biotechnologyforums.com/thread-1692.html [Accessed 28 Jul. 2017].
3. Kumar, P. (2016). Biotechnology: Introduction, Scope and Applications of Biotechnology. [online] Biology Discussion. Available at: http://www.biologydiscussion.com/biotechnology/biotechnology-introduction-scope-and-applications-of-biotechnology/11608 [Accessed 28 Jul. 2017].
4. Pai, J. (2003). Applications of microorganisms in food biotechnology. Indian Journal of Biotechnology, 2, pp. 382-386
5. Safety of Foods and Food Additives Developed by Biotechnology. (2017). [pdf] Available at: https://www.fda.gov/downloads/Food/GuidanceRegulation/GuidanceDocumentsRegulatoryInformation/IngredientsAdditivesGRASPackaging/UCM078765.pdf [Accessed 27 Jul. 2017].
6. Transparencymarketresearch.com. (2012). Animal Feed Additives Market - Global Industry Size, Share, Trends, Analysis, and Forecast 2011 - 2018. [online] Available at: http://www.transparencymarketresearch.com/animal-feed-and-feed-additives-market.html [Accessed 27 Jul. 2017].
7. Transparencymarketresearch.com. (2015). Food Additives Market by Type (Flavors and Enhancers, Sweeteners, Enzymes, Colorants, Emulsifiers, Food Preservatives, Fat Replacers) and By Source (Natural and Artificial) - Global Industry Analysis, Size, Share, Growth, Trends, and Forecast 2015 - 2021. [online] Available at: http://www.transparencymarketresearch.com/food-additives.html [Accessed 27 Jul. 2017].
8. Ukaegbu, G. ROLE OF BIOTECHNOLOGY IN FOOD PROCESSING. [online] Academia.edu. Available at: http://www.academia.edu/6586170/ROLE_OF_BIOTECHNOLOGY_IN_FOOD_PROCESSING [Accessed 28 Jul. 2017].

9. Wikipedia contributors. "List of food additives." Wikipedia, The Free Encyclopedia. Wikipedia, The Free Encyclopedia, 23 Jul. 2017. Web.27 Jul. 2017

CHAPTER 2

FOOD AND FEED ADDITIVES

"*Every time you eat or drink, you are either feeding disease or fighting it.*"

-Heather Morgan

WHAT ARE FOOD AND FEED ADDITIVES

Additives are the substances that are being added to the food directly during their processing to preserve the food for longer duration, and for stabilization of the food, these agents help in counteracting the undesirable properties and maintaining the general quality of food.

Food additives are the substances that are added purposely to foodstuffs to perform certain specific and technological functions, like giving specific colors to food, as a sweetening agent, and to preserve food for longer duration. According to the European union all the food additives are recognized by an E number on them and all the additives used in the food are embraced in the ingredient lists of foods. The product stamp must discover the function of the additive and specific substance used either by the name of the additive used or an appropriate E number. Commonly used additives that appear on food labels are antioxidants; they prevent deterioration caused by oxidation, other additives like emulsifiers, stabilizers, colors, gelling agents and thickeners, sweeteners and preservatives. On the other hand, Feed additives are the products that are used for animal nutrition for improving the quality of food from animal origin and enhancing the quality of feed. Unless the scientific evaluation is properly done assuring that the additives don't have harmful effects on human and animal health or on environment and proper authorization is given to them feed additives are not put in the market.

Food additives are broadly divided into two categories- Natural food additives and Synthetic food additives. Food additives that are made from natural sources like fruits, plant lard, animal lard, insects etc. are known as natural food additives. For example, vitamin C is a natural food preservative, while food additives made from artificial sources are known as synthetic food additives. For example, Butylated hydroxyanisole (BHA) and butylated hydroxytoluene (BHT) are the synthetic food preservatives. Synthetic food additives are mostly the flavoring agents. Flavoring agents in turn are again of two types- Natural flavoring agents and Artificial flavoring agents. Natural flavoring agents are the necessary oil, oleoresin, protein hydrolysate, essence or extract, or any other product of heating, roasting or enzymolysis in which flavoring agent is present that is derived from fruit or fruit juices, spices, vegetable or vegetable juices, herb, bark, edible yeast, root, leaf or other plant material, dairy products, eggs, poultry, meat, seafood, or fermentation products which are mainly responsible for use as a flavoring agent rather than giving any nutritional value to food. Conversely, the flavors derived from the ingredients that don't match the above description and are synthetically made are known as artificial flavoring agents. However, to ensure the well-being, both the food additives i.e. natural and synthetic food additives comes only from the group of substances that comes under GRAS (generally recognized as safe) list.

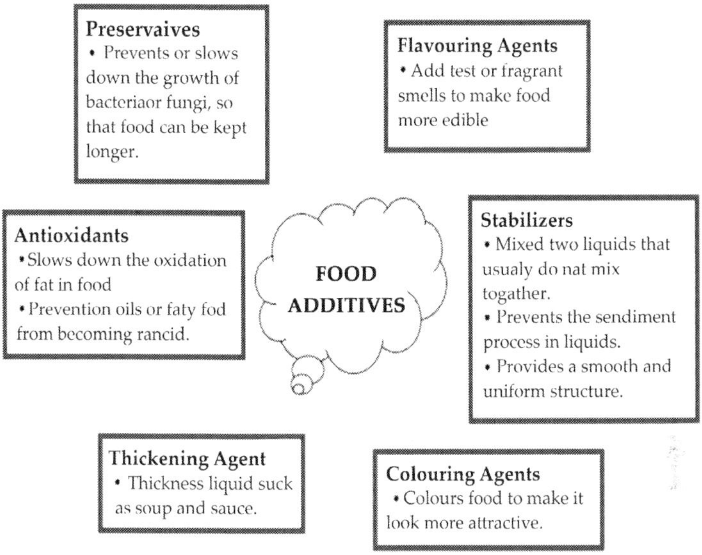

Figure 2.1: Some basic food additives

Source: http://my-chem-assignment.blogspot.in/2011/12/food-additives.html

All the food additives used in different food on the GRAS list-

- The food additives are approved by the FDA (food and drug administration), which ensures the safety and effectives of the additive used by the agency and the agency is satisfied with the additive they have added to the food.
- Food additives must be added to a specific amount only to ensure that it doesn't degrade the original quality of food.
- Food additives used in the food must satisfies the basic specific needs in food products, like protection against molds etc.
- Food additive must be such that it maintains the freshness and safety of food and must perform effective function for which it is used.
- Food additives must ensure that they are accurately listed on the label.

Feed additives as mentioned above are the food supplements that are given to the farm animals that don't get enough nutrients from the regular means and includes vitamins, minerals and amino acids. The diet of humans and animals contain a broad diversity of additives. The additives in the poultry diets, are primarily comprised to improve the efficiency or efficacy of the bird's growth or laying capacity of bird, improve feed utilization and in prevention of disease. The additives that are used in feed should first be approved for use and then they must be used as per the direction given with respect to duration of feeding and inclusion levels. Feed additives are specific for the age and type of birds. The guidelines for the usage of additives in feed are maintained by Australian government department of agriculture, fisheries and forestry. Some of the common feed additives that are used in poultry diets are- antimicrobials, emulsifiers, binders, antioxidants, pH control agents and various enzymes.

Some other additives like artificial and nutritive sweeteners, lubricants etc. are also used. There are dozens of companies which manufacture and distribute specific additives which each class of additives. Again, the ingredients and additives that are added must be noted on the label with their use and inclusion level meeting the standards as defined by the law. In some cases, additives are directly adjoined with the animal's diet in order to improve their value for human consumption, but in most cases, this is accomplished by the use of natural ingredients which contain significantly elevated levels of these nutrients that are deposited straight into meat and eggs.

Figure 2.2: Different types of feeds present in various fields

Source: http://www.toshniwalindia.com/animal-feed/

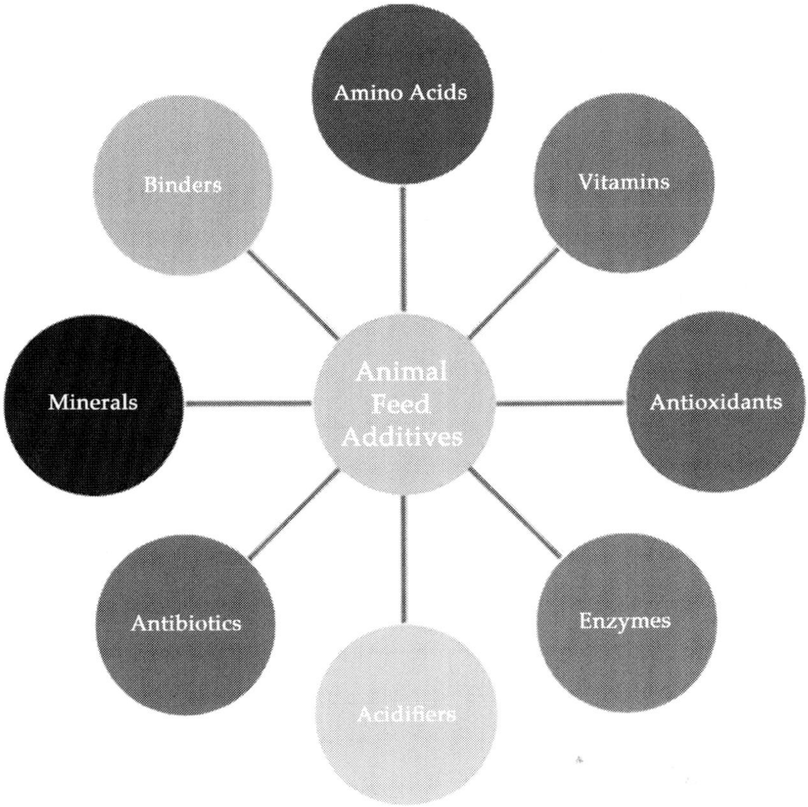

Figure 2.3: Categories of animal feed additive

Source: https://www.alliedmarketresearch.com/animal-feed-additives-market

Nutrient additives

Nutrient additives, are the substances which are added to increase the nutritional value of food and in turn clearly benefit the food. The objective of these additives is to deliver nutrients in great amount such that, the consumers get maximum benefit by consuming these nutrition rich foods. Some of the examples of nutrient additives used in food are:

- Vitamin D, which is added effectively to all milk sold in the United States,
- Vitamin B, is added to most U.S bread and grain products, with other essential minerals like iron etc. to substitute the lost

occurred when whole grains are milled into white flour for white bread,

- Vitamin C, which as antioxidant and antimicrobial properties is used as a useful preservative.

Being an antimicrobial and antioxidant agent, the main function of vitamin C is to slow down the spoilage of food and preventing destructive chemical reactions.

Due to these reasons, American food packagers add vitamin C (iso-ascorbic acid and sodium ascorbate) to bacon and other meat products to avert the development of potentially cancer-causing compounds.

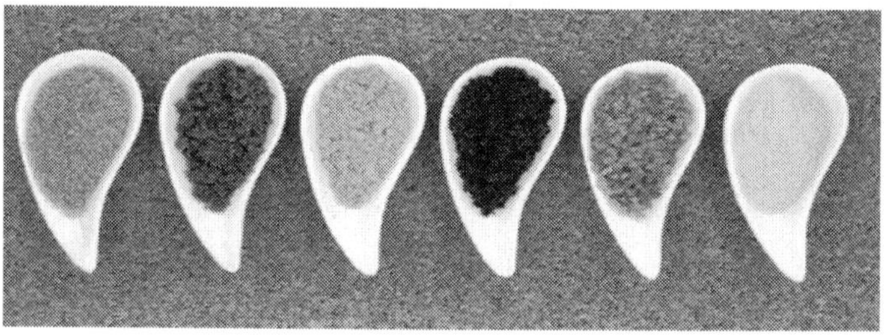

Figure 2.4: Some of the common nutrient additives

Source: https://www.maximumyield.com/common-nutrient-additives/2/1327

Color additives, flavoring agents and flavor enhancers

Color additives, flavoring agents, flavor enhancers they all make food look and taste good. Like the other food additives talked about, color additives are also of two types-

- Natural color additives and
- Synthetic color additives.

Natural color additives such as beta carotene, which is a yellow color pigment that is extracted from fruits and vegetables and are used to turn naturally white margarine into buttery yellow.

Other natural agents are annatto, which is a slight yellow-to-pink pigment obtained from tropical tree; chlorophyll, which is a green pigment obtained from green plants; carmine, which is a reddish extract of cochineal (a crushed leaf pigment); saffron, a yellow herb; and turmeric, a yellow spice.

Synthetic color additives like FD&C blue no.1, which is a blue color pigment that is made from coal tar and is used in soft-drinks, hair dyes, gelatin and face powders etc.

As discovered by scientists, the effects of coal-tar dyes are carcinogenic that are almost banned in many countries to be used in food, but they are still used in cosmetic products.

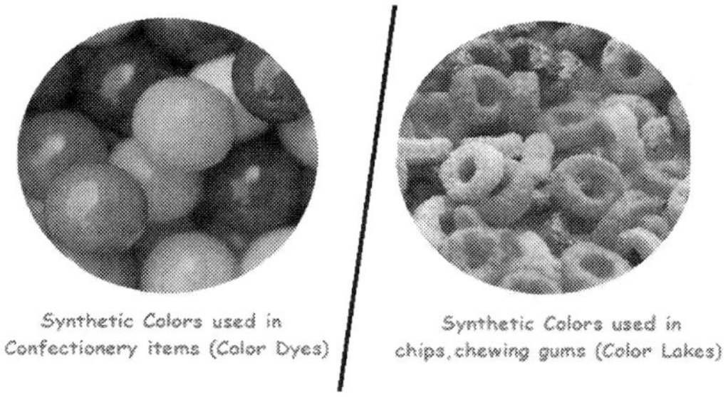

Figure 2.5: Colored food stuffs

Source: http://www.foodsciencesblog.com/synthetic-food-colors/

Flavoring agents, like salt, sugar, wine, vinegar and fruit juices are all examples of natural flavors, whereas, a teaspoon of fresh lemon juice in batter lends of cheesecakes gives it a special taste, an artificial sweetener, saccharin, is added to the morning coffee to improve its taste. All the natural flavors are reproduced by artificial flavoring agents.

The flavor enhancers are a little dissimilar kettle of fish. They boost the natural flavor of food instead of adding any new flavor to them. For example, monosodium glutamate (MSG), is the best flavor enhancer used in Asian foods to refine their flavors. But there are few cons of monosodium glutamate, it may cause mild reactions for short-term, like headache, sweating, facial numbness and tingling, flushing, and increase in the heart of people who are delicate to seasoning.

Spoilage of food is a total natural phenomenon which cannot be stopped completely, but this can be slower down to certain extent using food preservation techniques. Food spoilage includes- molding in bread, milk sours, rotting of meat and poultry, vegetables wilt, rancid fat etc. the former three spoilages of food are mainly caused by the microbes such as bacteria, mold, and yeasts, whereas the last two are caused when food is exposed to oxygen. Food preservation techniques like- cooking, canning, chilling, freezing and drying prevents their spoilage up to certain extent either by slowing down the growth of the microorganisms that live on the food or by shielding the food from the oxygen effects.

- Antimicrobials, are the natural or synthetic food preservatives that safeguard the food by slowing the growth of molds, yeasts and bacteria.
- Antioxidants, are also the natural or synthetic food preservatives that shield the food by preventing the food molecules to combine with oxygen.

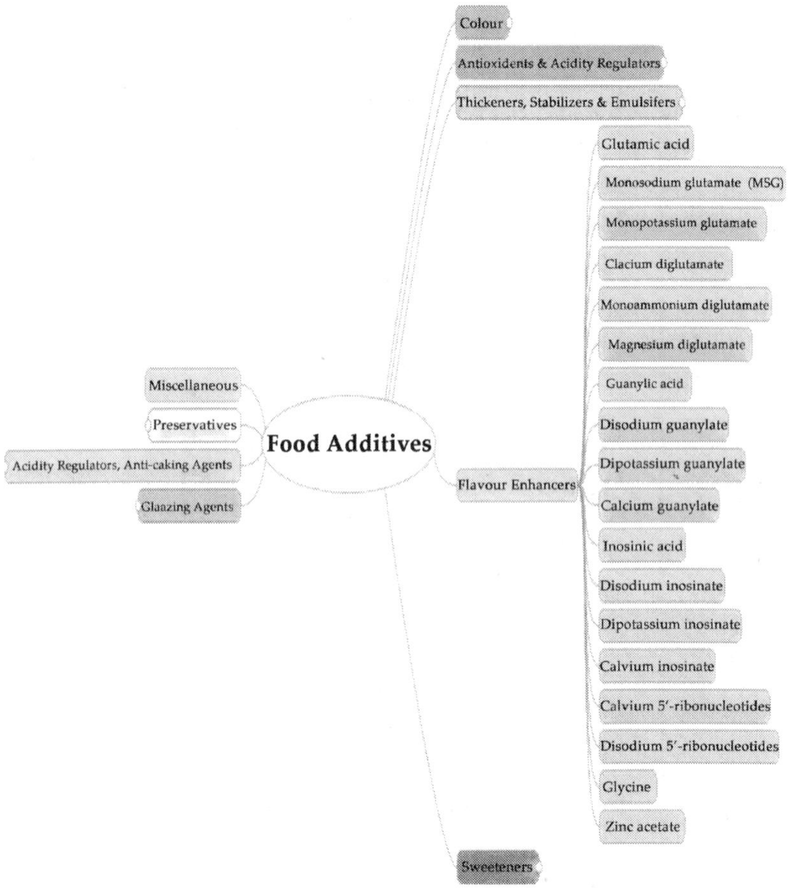

Figure 2.6: Broad classification of flavor enhancers used in food additive

Source: http://foodconstrued.com/2013/09/food-additives/

Table 2.1: Additives in Food

Additives	Found in ...
Ascorbic acid*	Sausages, luncheon meats
Benzoic acid	Beverages (soft drinks), ice cream, baked goods
BHA (butylated hydroxyanisole)	Potato chips and other foods
BHT (butylated hydroxytoluene)	Potato chips and other foods
Calcium propionate	Breads, processed cheese
Isoascorbate*	Luncheon meats and other foods

BACKGROUND OF FOOD AND FEED ADDITIVES

Since the earliest records of food preservation by humans, bioprocessing is being used to produce wide variety of food and feed ingredients. Microorganisms like molds, yeast and lactic acid bacteria are widely used for the conversion of raw food into a plethora of fermented products. It is because of the development of bioprocess technology that it has become possible to produce food and feed ingredients or bioprocessing aids. Lactobacillus species are the most beneficial microorganisms because of their long history of use. These are the first microorganisms that are used by man for processing food stuffs (Konigs et al., 2000) and for preserving the food by hindering invasion caused by other microorganisms that leads to foodborne illness or food spoilage (Adams, 1999). Because of the beneficial properties of Lactobacillus, they are increasing used in animal feed as they have the potential to replace antibiotic growth promoters. The self-care health industries and dairy industries are also actively promoting the use of Lactobacilli in foodstuffs.

During the year between 1970 to 1990, most of the articles were mainly focused on functions of food preservation and production. Many new products that contains Lactobacillus have been released in the market. These new products contain new Lactobacillus species of non-dairy origin, for example Lactobacillus rhamnosus, isolated from healthy human intestinal flora (Goldin et al., 1992).

Although, the concept of food additives is not new, the ancient Chinese used traces of propylene and ethylene unknowingly from burning paraffin to ripen fruit, originally additives were used to simply preserve the food. For preservation of food like meat or fish, smoking is used, which is an ancient method for preservation of food. This method relies partly on drying and partly on preserving chemicals in the smoke. The resins that are added in the Greek wine act as a mild preservative.

The first food additive that has been deliberately used since time immemorial was salt. It has been used in fish since ancient times. Romans made the fish sauce, liquemen in which the quality and fermentation process was controlled by addition of salt. Salt is a simple inorganic chemical and has been in use lavishly as a food preservative. The use of salt has become so familiar to people and due to its usage in great proportion people confuse it as an ingredient rather than a food additive, same is the sugar with sugar, it is also considered as an ingredient, in contrast to additive.

As mentioned above, salt is a simple inorganic chemical whereas, sugar is a more complex organic molecule. The white granulated sugar that is used for food preservation is a very pure chemical of great value which is important both in home or factory. Sugar acts as a functional additive in different types of foodstuffs, for example in some food products like honey and dried fruits, it acts as a preservative, whereas, in jams cane sugar is used as a flavoring agent. The only sugar compound that is not confused as an ingredient added to the food is saltpetre (potassium nitrate).

As discussed earlier, fermentation is an empirical method for preservation of food products. Alcoholic fermentation is itself a form of preservation. Alcohol, in the form of brandy, is used as an additive in brandied wines. Another historical example for in situ preservation by fermentation is production of lactic acid, as in cheese, yogurt, certain cured sausages etc. Romans also used the acetic acid of vinegar as a food preservation for fruits and vegetables. To prolong shelf life of fruits and vegetables, a modern method is the surface treatment of meat with acetic acid.

Another example of preservative in food that are today confused and classified as ingredients are spices and hops, though spices are used as a mild food preservative, still they are used as flavoring agents in food since ancient times. Hop resins, also function as preservatives, around 600 years ago, hops were added to the beer which was traditionally not appreciated or accepted, but gradually it was noticed that hopped beer's shelf life is more than the unhoped ones.

Figure 2.7: Spices used as traditional mild food preservative

Source: https://www.naturalproductsinsider.com/blogs/food-beverage-perspectives/2015/09/are-natural-antioxidants-the-food-preservatives-o.aspx

Traditionally, preservation and addition of flavor to the food was more important than adding any color to it. Food coloring gained importance during the medieval times, but it was restricted to the kitchens of the castles and great houses where chefs used too many colors of plant origin. Few of them includes- indigo, sanders from sandalwood, saffron from crocuses, alkanet from the roots of borage, greens came from the chlorophyll of spinach and sage, and pinks, blues, violets, and mauves from the petal extracts. Saffron, was used as a coloring agent in food in ancient Egypt, similarly, the upper-class Romans prefer to have white bread, which was prepared by sieving the flour and then adding alum to it, this practice was continued till the end of nineteenth century.

Later, in the 17th century, annatto derived from the seeds of the shrub Bixa Orellana and cochineal derived from the dried insect Coccus cacti, reached to Europe from the new world. The former was used to color cheese and bread till eighteenth century, after which saffron was used to color butter. There was no existence of food industry during that time and use of additives was only limited to the kitchens and to farmhouse and village technology. Since there was no food processing industry during the traditional period, there were no boiler water additives, sanitizers, processing aids, nor any agricultural chemicals.

In the early world, the food was cooked and stored in the lead vessels, leaded brass and bronze, and pewter and was served on pewter and lead dishes and plates. Romans used lead for collection and distributing the water and used lead pipes for water reticulation in many houses which is used even now in some old house of Britain and Europe. Romans used grape juice concentrated by boiling in lead vessels as they lacked sugar to be used as a sweetener. Lead was used because extra sweetness was derived from lead salts. Due to high consumption of lead content in food and soft drinks especially wine, that it led to lead poisoning and it became the major cause of decline of roman empire. The disease caused by lead poisoning in upper class Romans was named as "Devonshire colic". Early in 1905, when Australian state of Victoria food regulations were introduced, the first condition that was included in the food regulation was prohibition of lead in or in contact with the food.

To regulate the food additives that are added to food, the attempts of government had a very long history. Late, in the middle ages, saffron was used by pastry cooks to replicate eggs, and during seventeenth century, annatto was used for the coloring of butter. However, the use of both saffron and annatto was prohibited by French (in 1574) and Amsterdam (in 1641) respectively. In the eighteenth century in England, it was claimed that chalk and ground bones were added to whiten its color, however, it was later refuted. But undoubtedly alum was still used in this capacity. Some of the other additives that were used during that time are-calcium carbonate in the form of chalk, marble, or shells were added to cider and beer to countervail acidity; bean flour was used as a clarifier; and copperas (iron sulfate) was added for head production.

During the period of industrial revolution with the emergence of pioneer factories, dramatically, the adulteration rate was also increased due to the usage of cheaper substitutes and also because of the involvement of additives such as colors, thickeners and flavors, . In the mid-century, lead chromate, prussian blue, alum, black lead, copper carbonate, red lead, copper arsenite, and vermilion were all used in food and confectionery. Early in the 1840s, sodium nitrate was used for the preservation of butter. Form the 1830s to 1870s, fewer attempts were made in Australia for the preservation of meat with the help of phosphoric and lactic acids, sulfites, nitrates and nitrites. Later, emerged the salicylic acid and boric acid, which were used as flavored preservatives for milk, sausages, and cordials. In 1905, adulteration act was passed by the British Medical Association in Toronto, Canada. Under this act, the first list of food additives was prohibited for use was published. Another report by Victorian Government Analyst, Australia, covered colors, flavors, preservatives, alum, and saccharin. This resulted in aforementioned Pure Food Act (Farrer in 1980).

In the beginning of twentieth century, the accidental discovery of mauveine by William Perkin (1856), made available a large number of synthetic colors, few of which were utilized in food. Due to their tinctorial properties they are used in very small amount to acquire the desired results, and thus they are recognized as much safer additives than lead, copper, and even the arsenic salts which were in use for long. In 1906, United States of America, register 7 colors authorized to be used in foods and due to commercial rationalization, the number of dyes were speedily restricted elsewhere by manufacturers. During the second World War, the American record was expanded, and the list of colors authorized to be used in foods were acquired by several other countries. Also, in the early twentieth century, borates and salicylates were abandoned to be used as preservatives, and in their place benzoates and sulfites were used, regular flour along with corn flour, gelatin and gums, was employed both for thickening and stabilizing, and natural lecithin of yolk sac was used as a common emulsifier. After World War II, the changes came when carcinogenicity of additives and ingredients became an issue.

IMPORTANCE OF FOOD AND FEED ADDITIVES

Additives are the substances or the mixture of substances that are added to the food so as to prevent the food from early food spoilage. This can be done by adding certain substances to the food, other than the basic foodstuffs, to meet the aspects of production, processing, storage or packaging that is required. An additive that is been added may be reactive or non-reactive; it may or may not be nutritive; it should neither be toxic nor hazardous. Additives added to the food are been developed over several years to meet the needs of food production, as it is very different and difficult to make food on a large scale than making food in the kitchen at home. The main objective of additives is to ensure that the processed food remains in good condition throughout their journey from the factory to the shop and from shop to the consumers. Some of the additives are so essential that they are even used in organic foods. According to Food Standards Agency estimate there are around 850,000 cases of food poisoning reported annually in United Kingdom and microbial spoilage is the main reason for high levels of waste as food is transported from the farm to the table. Preservatives play a very important role to counteract the adverse effects of food poisoning. Without preservatives, this figure would be much more than recorded. The significance of sustainable food production has become well known only in the recent years, and attempts are made to minimize land and water requirements, and also the amount of energy that is being used to transport and store foodstuffs. According to United Nation Environment Programme report more than half of the food produced in the world is lost, discarded or wasted due to various reasons. This makes clear that it is important to preserve the food that is produced in a better way. Various forms of preservatives, stabilizers and antioxidants are all used for this purpose to keep the food in good condition until they are consumed. There are many kinds of food additives, that are designed to do an important job in making the food safer and more appealing.

Some of the chief importance of food additives includes provided by FDA-

- To improve flavor or impart desired color
- To control acidity and alkalinity
- To provide leavening
- To maintain the consistency of product
- To enhance or maintain nutritional value
- To support palatability and wholesomeness.

Other important features of additives are-

- They elevate the nutritive value of food. Additives such as amino acids, vitamins, minerals etc. are used to increase the nutritive value of food. A particular diet is required for the utilisation of thickening agents, emulsifiers, sweeteners, etc.
- They are required to add sensory value to the food. Color, taste, odor and consistency or texture, they are important for the sensory value of food but they may decrease during processing and storage. This loss can be cope using additives such as pigment aroma compounds or flavor elements, these compounds correct and readjust the decrease caused during storage or processing. Antioxidants helps to suppress the development of "off flavor" derives from fat or oil oxidation.
- They are important to increase the shelf-life of food. Today, people are more relied on processed foods than the normal food, due to this the current forms of food production and distribution demand for longer shelf life to the food. In addition to that, the world supply chain requires preservation by avoiding deterioration as much as possible. By extending the shelf life of food it can be protected against microbial spoilage. For instance, use of antimicrobial additives and the active agents used can suppress and retard undesired chemicals and physical changes in food. The physical changes can be attained by stabilizing the pH by using buffering additives or by stabilizing the texture with thickening and gelling agents which are polysaccharides.

It should be completely understood that the food additives and their degradation level must be non-toxic at their recommended levels of use. This is also applied uniformly to acute and chronic toxicity, particularly the teratogenic, mutagenic, carcinogenic and mutagenic effects.

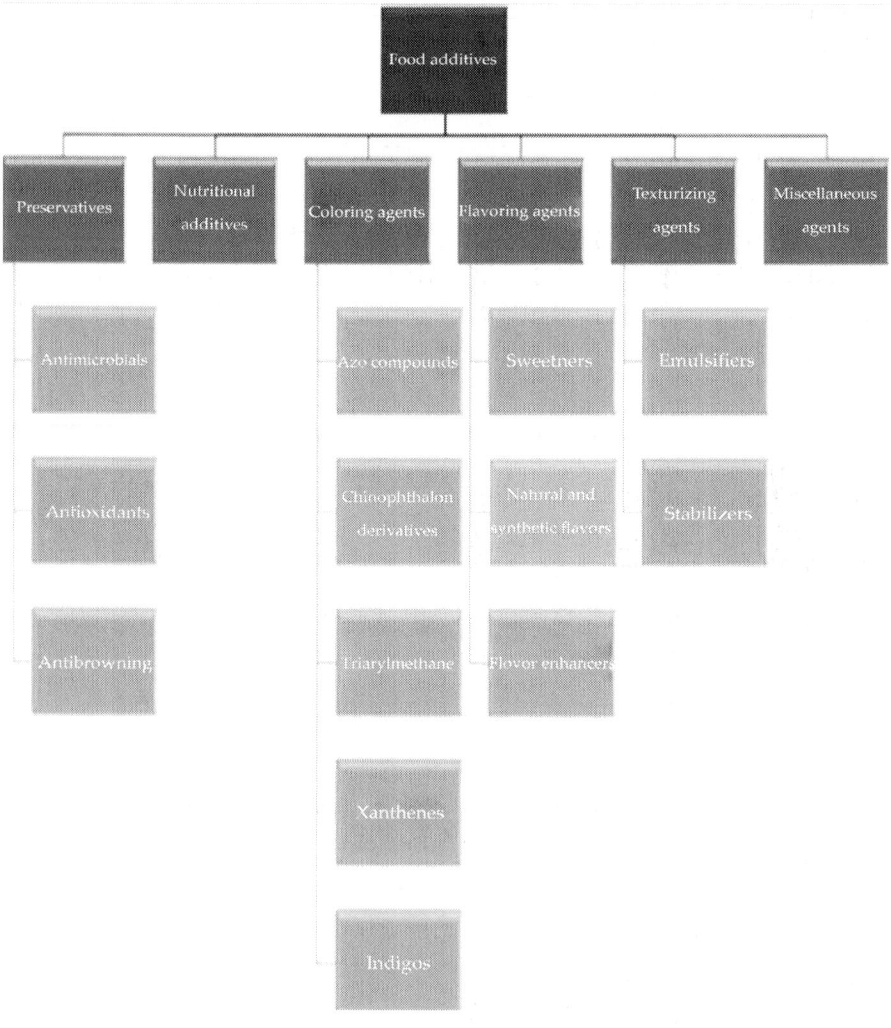

Figure 2.8: Groups and sub-groups of food additives

Source: https://www.researchgate.net/publication/264808591_Adding_Molecules_to_Food_Pros_and_Cons_A_Review_on_Synthetic_and_Natural_Food_Additives

The necessity of food and feed additives

As discussed above, people have been using food additives for thousands of years. Use of salt for the preservation of meat is a one such example. The necessity of food additives increases exponentially with the accelerated reliance on packaged and stored foods that needs to provide a constant supply of different and plentiful food. Since, people are much more dependent on processed foods, it should be thoroughly understood that the processed foods cannot exist without additives. The main advantage of additives is that they keep the foods from going bad on the shelf. Additives protect the foods so they remain safe to eat despite the threat from dangerous microorganisms. Preservatives protects the food against microbial decay, anti-oxidants, sequestering agents, emulsifying and stabilizing agents and buffering agents etc.

Role of food and feed additives

The five main roles served by food additives are:

Maintaining the product consistency-

Emulsifiers are used to provide consistent texture and prevent the products from separating. To give uniform texture stabilizers and thickeners are used, and to enable the substances to flow freely, anti-caking agents are used.

Prolong the palatability and wholesomeness of food-

Most of the food products get contaminated due to the various microorganisms they are exposed to, this causes food-borne illness to the population. Use of preservative reduce the spoilage that air, fungi, bacteria and yeast can cause. Preservatives like anti-oxidants assists the baked goods to preserve their flavors by preventing the oils and fats from becoming rancid. Also, they keep the fresh fruits from turning brown when they are introduced or exposed to air.

Improve or preserve the nutritional value of food-

Due to protection and enhancement of foods, it has become possible to improve the nutritional status/value of population. For instance, minerals and vitamins are added to numerous food including cereal, flour, milk, and margarine. The use of vitamins and minerals as a food additive helps to make up their low levels in individual's diet by enhancing their nutritive value.

To provide leavening, and control the acidity and alkalinity-

Leavening agents when heated they release acids and react with baking soda to help biscuits, cakes, and other baked goods rise. Specific additives help in the adjustment of acidity and alkalinity of foods to gain wished color, taste or flavor.

Providing color and improving flavor-

There are certain colors that are used to enhance the appearance of foods. Many spices and natural or synthetic flavors are used to provide the best flavor to the food. Food additives sometimes have more than one use, few examples of certain food additives with their common roles according to their classes are listed below-

Substances used to restore or add color/flavor to the food-

Coloring retention agents are used to intensify the color of food. To protect the food against deteriorations caused by microorganisms, preservatives are used.

Artificial sweeteners are used to bring sweet taste in food for fewer kilojoules or calories than sugar. Flavor enhancers are in use from long back to increase or improve the flavor or aroma of food. Flavoring is necessary to restore the taste loss caused by processing, maintain uniformity and helps in making the food more palatable.

- Anticaking substances are used to keep powdered products like salt, etc. flow freely when poured.
- To prevent the oil and water mixture from separating into layers, emulsifiers are used.
- Food acids aids to prolong a constant bitterness in food.
- Humectants prevents dried fruits from drying out.
- To improve the texture of foods like processed meats, etc., mineral salts are used.
- To upgrade the texture and preserve the uniform consistency, thickeners and vegetable gums are used.
- Stabilizers are used to keep the uniform dispersion of substances in food.

- Flour treatment agents are being in use to boost the baking quality and appearance of flour.
- To give shiny appearance or to provide a protective coating to a food, glazing agents are used.
- Propellants are used to propel food from a container. Propellants are the chemical substances that are used to produce energy or pressurized gases.

Additive	Molecule
Colours	Anthocyanin, Betaline, Carotenoids, Curcumin
Sweeteners	Sucrose, fructose, stevioside, Thaumatin
Flavour	Capsicum, garlic, vanilla, celery
Emulsifiers	Fatty Acid Esters of Glycerol cetearyl alcohol, polysorbate
Gelling agents	Natural gums, starches, pectin, agar-agar
Preservatives	Various antioxidants
Stabilisers	Agar or pectin
Thickeners	Polysaccharides and proteins

Figure 2.9: The above picture table depicts the additives and molecules

Source: http://www.laboratory-journal.com/science/umwelt/future-plant-bio-technology-application-foods-and-ingredients

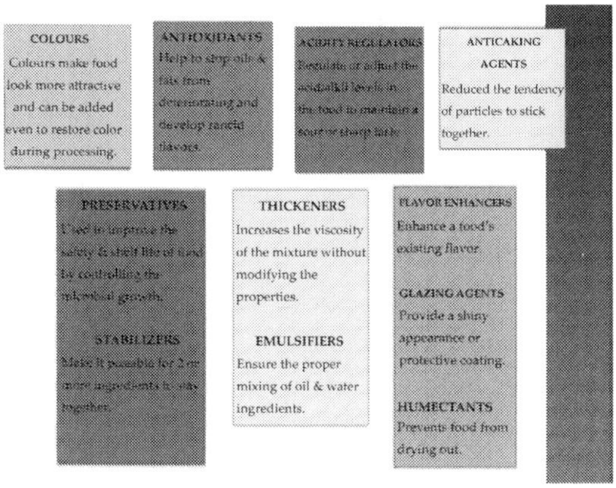

Figure 2.10: Role of various additives

Source: https://www.slideshare.net/mokshacb/food-additives-market-survey

ARTIFICIAL COLORINGS-

Majority of the artificial coloring agents are the synthetic chemicals that do not exist in nature. Colorings are merely used in all foods with low nutritional value such as candy, soda pop, gelatin desserts, etc.; therefore, one should completely avoid all foods containing artificial colors.

Table 2.2: Uses and risks of artificial coloring agents

S.No.	Artificial Coloring Agents	Used in coloring of	Possible Risks
1.	BLUE 1	Beverages, Candy, and Baked goods	Inadequately tested; hint/ suggestion of small cancer risk
2.	BLUE 2	Pet food, candy, and beverages	This artificial color is suspected to cause brain tumour in male mice, but it is not yet proved. FDA certified it as food with no reasonable certainty of harm as of now.
3.	CITRUS RED 2	Skin of some Florida oranges	Studies demonstrate that this additive leads to cancer, but there is no risk except eating the peel.
4.	GREEN 3	Beverages and Candy	According to 1981 study, this dye gave hints of bladder cancer, but later on FDA re-analysed the data and reached to the conclusion that the dye is safe. Although, this carcinogenic dye is rarely used in coloring the foodstuffs.

5.	RED 3	Cherries in fruit cocktail, baked goods and candies	According to 1983, review committee report requested by FDA, there was evidence of this dye causing thyroid tumours in rats was recorded.
6.	RED 40	Gelatin desserts, soda pop, pastry, sausage, and pet food	It is the most tested and widely used dye in foods. FDA review committee recognized complications, but concluded that the evidence of harm was not consistent or substantial. RED 40 is majorly used in junk foods.
7.	YELLOW 5	Candy, gelatin desserts, baked goods, and pet food.	This is the second most outspread dye after RED 40, the potential cause of this dye is mild allergic reaction, primarily in aspirin-sensitive patients.
8.	YELLOW 6	Sausage, baked goods, beverages, candy, and gelatin	The third most widely used artificial coloring agent, this dye leads to tumours of adrenal gland and kidney. However, according to FDA after reviewing the data, found that it does not pose any significant cancer risk to humans. YELLOW 6 may cause momentary allergic reactions.

ARTIFICIAL AND NATURAL FLAVORING AGENTS-

Thousands of chemicals are used to give colors to the foods that mimic the natural flavors. Most of the chemicals are used in single flavoring, for example cherry soda pop. Most of the flavoring agents or chemicals occur in nature and are probably very safe to use, but are predominantly used in junk foods. Almost all the companies keep the individuality of natural or artificial flavorings a deep secret. Some of the flavors used may also cause problems to persons who show sensitivity to certain flavors, i.e. MSG or HVP.

Cyclamate-

This high-potency sweetener is used in diet foods until 1970, after which it was banned. Studies on animals indicate that this sweetener leads to cancer, later, it was realized that it does not lead to cancer directly, but it influences the potency of other carcinogens to harm the testes.

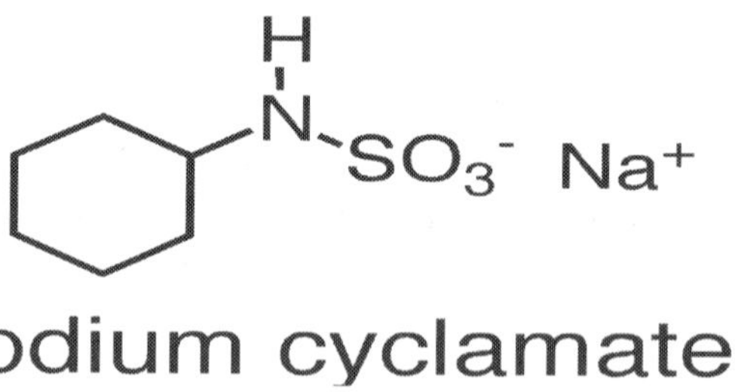

Figure 2.11: Chemical structure of sodium cyclamate

Source: http://www.sweetenerbook.com/cyclamate_2.html

Artificial Sweetener-

Used in diet products, packets and fountain drinks. Saccharin, which is 350 times sweeter than sugar, is used in diet foods or as tabletop sugar substitutes. Animal studies shows that saccharin leads to cancer of the urinary bladder. Other studies on rodents, manifests saccharin causes cancer of uterus, skin, blood vessels, ovaries, and several other organs. Many other studies, exhibit that saccharin elevates the strength of other cancer-causing chemicals. According to the epidemiology study by National Cancer Institute, it was found that the artificial sweeteners like saccharin and cyclamate are associated with the higher prevalence of bladder cancer.

In 1977, FDA banned the use of saccharin in foodstuffs, because the studies showed that it causes cancer in animals. However, Congress involved and authorize its use with a warning notice labelled on the food products. Saccharin was further replaced by aspartame (NutraSweet) in many food products. Later, in 1997, the diet-food industry started pressuring the Canadian and U.S governments and the World Health Organization to remove saccharin from their list of cancer causing chemicals, as saccharin causes bladder cancer in male rats, but mechanisms of those tumors are so different that it would never cause cancer to humans.

In May 2000, U.S. Department of Health and Human Services finally removed saccharin from the list of cancer causing chemicals. This action perhaps results in removal of warning notice labelled on food products containing saccharin, increases its usage in soft drinks and other foods.

Nitrate-

It is used as a preservative, coloring, and flavoring agent in bacon, ham, smoked fish, luncheon meats, frankfurters, corned beef, etc. Meat processors used sodium nitrite as it stabilizes the red color in cured milk without which it would look gray, sodium nitrite not only acts as a stabilizer but it also gives a characteristic flavor to the food. It is recommended to use sodium nitrate in dry cured meat because nitrate slowly breaks down into nitrite. Adding nitrite directly to the food leads to the formation of small amounts of vigorous nitrosamines (cancer-causing chemicals), especially in fried bacon. Companies therefore, now prefer to add ascorbic acid or erythorbic acid to bacon to suppress nitrosamine formation, to reduce the problem. While nitrate and nitrite are not very risky chemicals, still it is better to avoid their usage up to an extent.

However, the meat industry claims that both nitrate and nitrite prevents the growth of bacteria causing- botulism poisoning. Nevertheless, U.S Department of Agriculture retort the statement saying that same could be done by freezing and refrigerating the meat, for this they developed a much safer method using bacteria that produces lactic acid. Since, the nitrite is mainly used in salty and fatty foods, nutrition- conscious consumers have strong reasons to avoid nitrite-preserved foods.

One of the biggest challenges that food faces before selling in the competitive market is ingestion of chemical food additives. Therefore, in order to overcome the biggest dangers to health, food manufactures must preserve the food for longer-shelf life, process the food properly, emulsifying it, stabilizing it, bleaching it, coloring it (so as to provide attractive ripe to food), adding flavors, spraying it with insecticides, with nematocides, with rodenticides, and fungicides, de-germ it, de-bran it, curing it, and etc. By the time food reaches to the mouth, it is loaded with plenty of chemicals, many of which are of high-toxicity and carcinogenic, few of them are insufficiently tested and their effects are still not known. Therefore, it is endorsed to use these additives to secured or safer amounts.

Potassium Bromate-

Potassium bromate is used as a flour improver in breads and rolls. Potassium bromate is used to expand the volume of bread and to produce the fine crumb structure of bread, most of the times bromate breaks down to form innocuous bromide. Bromate is proclaimed to cause cancer in animals, but the tiny amount of bromate left in bread pose a small risk to consumers. Nonetheless, use of potassium bromate has been banned worldwide except in countries like Japan, United States and rarely in California. In 1999, the Centre of Science un the Public Interest appealed FDA to ban the potassium bromate usage in foods.

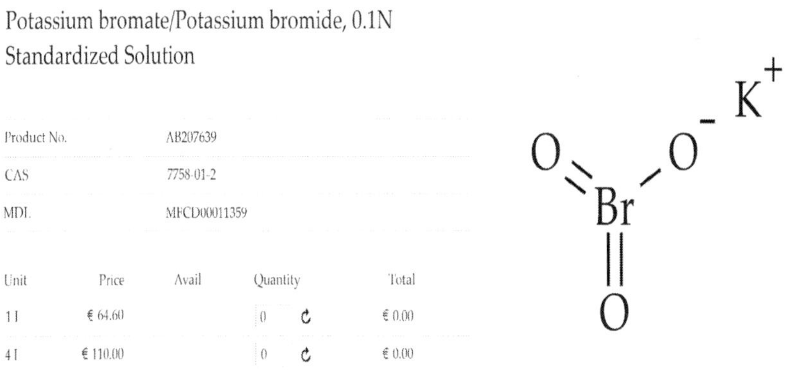

Figure 2.12: *Chemical structure of potassium bromate*

Source: https://www.abcr.de/shop/en/AB207639

Figure 2.13: Potassium bromate powder

Source: http://modernsurvivalblog.com/health/there-is-a-carcinogen-in-your-flour/

DEVELOPMENT AND APPROVAL PROCESS OF ADDITIVES-

The Food and Drug Administration (FDA), in the united states is assigned with the authority of overseeing the welfare of food and feed additives. But it was not same always, earlier there were no safety standards existed for food or food ingredients/additives at all. Later, in the 20th century, laws were passed to confirm the safety of meat, eggs and poultry. Currently, the system which assures the safety of additives, was put into place with the Food Additives Amendment act of 1958, for the safety of food, drug and cosmetics. In 1960, the Color Additive Amendment, this amendment ensures the burden of proof for the safety of additives completely on the food manufactures. FDA, now regularly monitors and evaluates the safety of the additives used. The food manufacturer appeals FDA for the demanding approval process, for which the manufacturer must have extensive evidence for the additive he is seeking approval for is safe for its intended usage. The FDA evaluates the safety of intended use and the amounts of the ingredient used, amount of additive a person is likely to consume, the possible consequences of consuming small amounts from several sources over time, and the potential effects of the additives on health. If the information submitted by the manufacturer for the additives used doesn't hold any safety concerns, then after reviewing all the information, FDA sets some limits to the product so as to ensure the safety. If the product is used in the labelled limits market on the product, it is intended to be protective until someone consumes the ingredient in higher amounts.

All the new and modern additives that are added to the food must undergo FDA-approval process. Furthermore, a group of almost 700 ingredients fall under the category of GRAS (Generally Recognized as Safe). These ingredients are marked to be safe by USDA or FDA, even before the 1958 food additive law. After the law was passed, based on their extensive history of safe usage in food they are listed in GRAS category. Some of the ingredients that fall under GRAS category are- salt, sugar, spices, monosodium glutamate (MSG) and caffeine. Although, the review process followed by GRAS is very different, but it is none less rigorous than for the new food additive used. To certify the ingredients to fall under GRAS, there must be a widely known safety information to the public and consensus among the qualified experts to ensure the safety of the ingredient. According to the former Acting Director of FDA Office of Food Additive Safety, Dr. Mitchell Cheeseman, "Both food additive and GRAS classifications emphasize safety for consumers and set a high standard that new ingredients and additives must meet to be used in foods and beverages."

The development process for additives and ingredients used in food commonly involves inputs from highly-qualified experts who reviews the ingredient's safety. Moreover, before getting authentication to be used in foods, additives undergo lots of scientific research and long-term studies in mice and rats, to see the effects of the additive or ingredients used is safe on health conditions like cancer, brain health, heart health reproductive health and more. After careful consideration and reviews, FDA permits the use of the additive or ingredient.

It almost takes several years for the manufacturer to assemble all the scientific evidence that are required by FDA to construct the safety of the ingredient or additive used. However, even after allowing the use of additive in food, FDA still holds the right to forbid its usage in food if the ingredient or additive currently used in foods is causing any harm to human health. FDA has the ability to demand the manufacturers for the additional data required including the safety level of the additive or ingredient used in the food or else it can restrict the use of ingredient until the concern is resolved.

Getting approval for the ingredient or additive added to the food had to cross a rigorous road once the welfare of an additive or ingredient has been established, the European Commission instigate the mandatory legislation for the specification of an additive used, and for its countenance in food, the process is commonly known as Food Approval. To ensure that the additive used has the same composition and purity as examined and assessed there must be requisite specification or description. These specifications are acknowledged by the technical experts of the state, under the chairmanship of the commission.

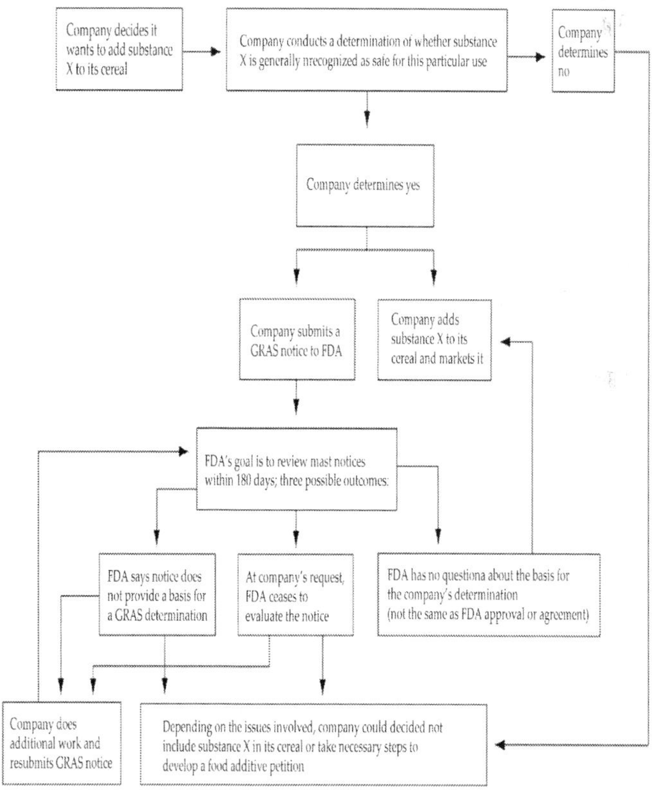

Figure 2.14: Flowchart depicting the complete approval process

Source: https://ehp.niehs.nih.gov/121-a126/

The judicial or legislative process for food approval is a complex one. It begins with the commission's proposal and approach, but for the end settlement National governments in the council of ministers and MEPs in the European Parliament jointly takes the decision. The final decision includes the EU-wide regulation that records the food in which additive is enabled and also its level of use. Once the additive qualifies its safety test and is legally authorized to be used in foods, it is given an E number, this not so famous E number guarantees the food purity and safety.

The Official Journal of the European Communities published a complete list of permitted additives, and the foods in which they are allowed, in their own legal journal. Additionally, individual specifications are also published in the Official Journal.

E-numbers are solely the code numbers that are used to recognise and discover the food additives that have been shown to be risk-free and are officially approved by EU for their usage in food.

E-numbers are simply the code numbers used to identify food additives that have been shown to be safe and officially approved for use in food across the EU. Each group of additives are allocated with specific blocks of numbers. For instance,

- The colors added to the food are all in the E100 series (example- E150 is caramel and E162 is beetroot red);
- -Preservatives belong to E200 series (example- E202 is potassium sorbate and E211 is sodium benzoate); --Antioxidants comes under E300 series etc. In all approximately 322 additives have been legally authorized to be used in food across the EU, which includes 42 colors and 17 sweeteners.

Figure 2.15: Various E-numbers

Source: http://www.what-is-eczema.com/types-of-food-additives.html

Figure 2.16: Some of the food additives and their E-numbers

Source: http://www.dailymail.co.uk/health/article-480207/Food-watchdog-condemned-totally-inadequate-response-harmful-food-additives.html

Acquiring a food approval is not a quick process, one must be patient enough to get approval for the new additive or ingredient used, as there is a hierarchy that is been followed and the additive are evaluated based on various tests and examinations. For example, in the year 2000, an application was made for the sanction of two new intense sweeteners-acesulfame/aspartame salt and sucralose. They completed the safety estimation by the year 2002, after which they were subjected to the judicial or legislative process to begin. The entire judicial process was finally done by the year 2003 and the two new sweeteners gained authentication to be used in foodstuffs in early 2004. This complete package almost took 3 years to finish.

Approval Process

For the approval of the new ingredient or additive, or new use of an already approved additive the manufacturer should appeal the FDA. To exhibit that the additive used in the food is safe for the proposed use in an animal food and for human consumption, the manufacturer submits a Food Additive Petition (FAP). The FAP must include the adequate information to demonstrate that the food is safe and also fulfils its proposed use, under the conditions that are detailed in the petition; once it has been exhibited, FDA can now issue a regulation which addresses the food additive and its use. The regulation specifies the amount of the additive used in an animal food, type of animal food that contains the food additive, and the demand for labelling the additive or animal food containing the additive. Once approved, an animal food additive must be used within the limited period of its validity.

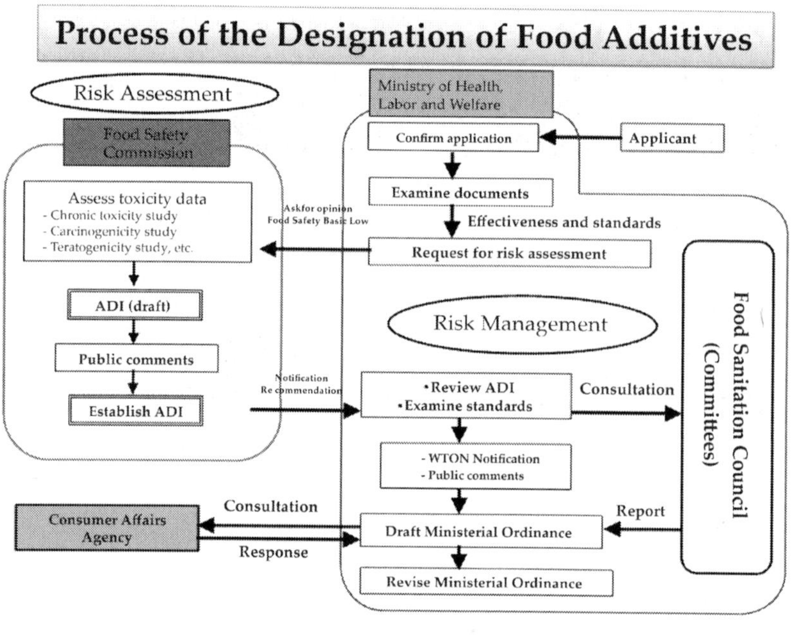

Figure 2.17: Process of designation of additives

Source: http://www.mhlw.go.jp/english/topics/foodsafety/foodadditives/

General information that are included in the FAP-

- Originality and arrangement of additive or ingredient used in the food, including its manufacturing method and controls;
- Expected usage and marked labelling such as shelf-life, warning, directions of use etc.;

- Data manifesting the nutritional, physical, or other technical effect;
- Various analytical methods for the additive and for animal foods comprising the additive;
- Recommended tolerance for the food additive;
- Planned regulation and environmental assessment or evaluation.

TYPES OF FOOD AND FEED ADDITIVES

Food Additives

Food additives have been used for centuries to enhance and protect its taste, texture, nutrition and also the appearance of food. According to Lauren Swann (Nutrition Consultant), food additives and preservatives are used to prevent foodborne illness, allow the passage of food to geographic areas that otherwise wouldn't be possible, and for the competent manufacture of products to constantly meet the conventional quality standards from batch to batch.

To enhance the texture of food and to prevent the mixtures from separating, food chemists use a variety of the following types of natural and chemical additives in food as discussed below-

- EMULSIFIERS- Emulsifiers are used to keep the two unfriendly liquids, like- oil and water together, such that the mixture of the two remains smooth. Emulsifiers, like- lecithin and polysorbate are used to keep the liquid-plus-solid, for example chocolate pudding, from dividing into liquid and solids. Emulsifiers are added to oil and water-based mixtures so they stay merged over the passage of the food product's shelf life. Other examples of emulsions in average foods include vinaigrette dressings, milk and mayonnaise. Lecithin from egg yolk and soybean are typically used emulsifiers in the food supply. Others comprises of diacetyl tartaric acid ester of monoglycerides (DATEM) and sodium stearoyl lactylate. These are frequently used in commercial bread doughs, artificial whipped creams and dried, liquid or frozen egg whites.

- STABILIZERS- Stabilizers, like alginates (alginic acid) obtained from seaweed, helps in making the food, like ice-cream feel richer, smoother and creamier.
- THICKENERS- Thickeners are nothing but the natural gums and starches, and their main function is to add body to foods. Example of thickeners include- apple, pectin or cornstarch.
- TEXTURIZERS- Texturizers are used to protect the food from turning soft and pulpy. For example, calcium chloride, which is used to keep foods like canned apples, tomatoes, or potatoes from turning mushy.
- pH CONTROL AGENTS- Ingredients that either retain or regulate the acidity or alkalinity of foods are known as pH control agents. Citric acid, acetic acid and sodium citrate are extensively used mediators and are often found in gelatins, jams, ice cream and candies. For cheese- making, lactic acid is used as an acidity regulator, and adipic acid can be found in bottled fruit-flavored drinks.
- ANTI-CAKING AGENTS- Anti-caking agents are added to crushed or powdered ingredients, example- powdered milks, egg mixes, sugar products, flours and baking mixtures to avert lumping, caking or sticking. There are numerous agents to elite from, including calcium phosphates, silicon dioxide, silicates (calcium, aluminium and tricalcium) and stearic acid.
- HUMECTANTS- Humectants helps in keeping the food moist. Examples of humectants includes glycerin, honey, sugar polyols (glycerol, sorbitol, xylitol, maltitol) and propylene glycol, which are found in candy, baked goods and salad dressings. It's essential to note that polyols are also used in foods as a low-calorie sweetening option, mainly for sugar-free chewing gums, candies and other low-calorie foods.
- LEAVENING AGENTS- They are incorporated into doughs and batters to rise the volume, shape and texture of baked goods. Some of the leavening agents include baking powder, beer, buttermilk, yeast, whey protein concentrates and yogurt. Leavening agents are used in a wide variety of sweet and savoury products, these can be found in cakes, cookies, breads, biscuits, scones, muffins, and soda bread.

Feed additives

Catalysts are proteins that empower particular substance responses. As the response is finished, the chemicals separate and wind up noticeably accessible to contribute in encourage responses. Although, the creatures and their related gut microflora produces numerous proteins, they are not primarily ready to deliver satisfactory amounts of particular compounds or create them at the precise positions to help ingestion of all segments in typical feedstuffs or to reduce against nutritious factors in sustain that limit assimilation.

A portion of the oat grains, for example, rye, grain, wheat, sorghum, have dissolvable long chains of sugar units that can catch colossal amounts of water amid processing and frame exceptionally gooey gut substance. Compounds, reaped from microbial maturation and further added to nourishes can upset these bonds amongst sugar and fundamentally reduce the gut content consistency. Lower consistency brings about advanced assimilation as there is further communication of the stomach related chemicals with nourishes and subsequently more total processing; better ingestion as there is better contact among the absorptive surface of the gut and the processed bolster supplements; and upgraded wellbeing as the dampness and supplement positions in the fertilizer are limited which additionally lessens the supplements accessible for unsafe gut microflora to duplicate and test the feathered creatures. For instance, necrotic elements, caused by Clostridium perfringens, this illness is a perpetual intestinal disease which brings about diminish execution and mortality. The business catalysts created decreases the negative impacts of phytates, which are the plant stockpiling wellspring of phosphorous, and additionally, it likewise ties different minerals, proteins, and vitality, which lessens their accessibility to the flying creature (i.e. insects). The examination is continuing for the advancement of proteins that are more compelling in supporting capacity under a more extensive scope of handling and stomach related conditions. The new chemicals created may incorporate those that are fit for diminishing poisons which are delivered amid bolster decay and empower assimilation of starches. Sustain added substances are quickly grasped by the poultry business and have facilitated the extension of critical new innovation to advance the utilization and availability of in-nourish catalysts.

The following are a portion of the cases of encourage added substances utilized

- ANTIOXIDANTS- There is a scope of wellsprings of receptive oxygen species in like manner digestion, notwithstanding those that comes specifically from encouraging fixings. Oxidative stress exasperates the ordinary cell work, harm tissues and direct well-being status. Hostile to oxidants fixed these atoms and decrease their harm. • ACIDIFIERS- Bolster acidifiers are supplementary to the encourage in order to bring down the pH of the nourish and in this way the gut condition. A lower pH is probably going to oblige or halfway point of confinement the development of pathogenic intestinal organisms. Acidifiers happen both as natural or inorganic acids or their related salts. They can utilize their antimicrobial activity both in the encourage and everywhere throughout the gut. Wellbeing and execution invigorating impacts have appeared for a various natural acid like-formic, citrus, propionic, fumaric, and lactic acids. In any case, the total advantages of natural acids unmistakably rely on the type of connected natural corrosive, which might be secured or unprotected, uncontrolled factors like buffering limit of fixings, the presence of whatever another microbial specialist, cleanliness of generation condition, at that point the heterogeneity of microorganisms. • PROBIOTICS-Probiotics are the live mono or blended culture of microorganisms that are non-pathogenic, impervious to gastric and bile acids, and when they are ingested, they can usefully influence the host individual by enhancing the components of intestinal microbiota. The key recommended methods of activity of probiotics include:

  Adversarial activity headed towards the microscopic organisms that regularly causes maladies by the production of bacteriocins, hydrogen peroxide and natural acids that hinder their growth Aggressive avoidance which means rivalry for destinations to connect to the intestinal mucous layers and along these lines pathogenic smaller scale life forms are restricted from involving the stomach related tract.

  Battle for nutritious substances. Probiotics have additionally been depicted to indicate immunomodulatory properties commonly through impact of gut microbiota creation and in this way influencing commonly the intrinsic and versatile invulnerability. In that way, they construct conditions in the digestive system which support advantageous microbes and thwart the improvement of pathogenic microscopic organisms.

- PREBIOTICS-Prebiotics are the inedible sustenance fixings which energize the development or activity of a specific number of microbes in the gastrointestinal tract of host creature. At the point when prebiotics move into the gut, they fill in as a substrate for the endogenous steady microorganisms in this way can fortify aggressive end of pathogenic microorganisms and specific colonization by helpful microorganisms. Prebiotics, as mannan-oligosaccharide (MOS) fructo-oligosaccharide (FOS) and galacto-oligosaccharide (GOS) are tried in poultry.

  The proposed component of activities of prebiotics incorporates: Bringing down the pH of the gastrointestinal tract through lactic corrosive creation; deterring the settlement of pathogens, producing systemic outcome on incitement of insusceptible reactions.

BENEFITS AND PROBLEMS OF PRESERVATIVES

Preservatives nearly globally catch a bad rap. But there's certainly a give and take with preservatives. Few of them are truly helpful and expand the quality of the food that is consumed by people for extended periods of time.

There is still a very large market for preservative-free foods, though, not all preservatives should be considered as evil. There are some pros and cons with everything that comes up, so is with the preservatives that are used in foods, benefits, and problems of some of the preservatives are listed below:

Benefits of Preservatives

Some of the natural preservatives that benefits the food in a good way are listed below-

- Drying

Dehydration is of the best methods used to preserve foods. For instance, in fruits and veggies, dehydrating takes out the moisture and delays the product's capacity to create bacteria and to spoil, while concentrating the nutrients, and also the flavor. Conclusion is: the food will remain healthy and tasty for several years when stored properly.

- Pickling

Pickling foods is the method which is in use since long for the preservation of foods. Pickling is done by using a solution of vinegar, salt and few other spices that preserves the picked food as well as adds lots of flavors to such foods. This is accomplished by solution of vinegar, salt and other spices that not only protects the pickled food, but also enhance whole lot of flavor in it. Pickled vegetables and meats delivers lot of flavors without worrying about the used methods to add calories.

- Salting

Salt is the rival of all the bacteria that are present. Nearly all microorganisms that are attracted to the food will not be able to boom in the existence of salt. Now, since the people are more concerned about their sodium levels, foods such as salted cod and country ham aren't actually preferred. but, even if they are used they won't be able to create any health problems to consumers due to the presence of unnatural chemicals that are used in preservation of food.

Problems of Preservatives

Use of chemical enchanters or the artificial preservatives in foods leads to variety of problems to consumers. Some of the most potential dangers to human health are listed below-

- Nitrates

Sodium nitrates are used to preserve numerous foods, for example-hotdogs, sausages and other cured foods. The objective of nitrates is to break off the bacteria that causes botulism, but according to the National Toxicology Program (NTP), conducted by U.S., diseases such as pancreatic cancer and lung cancer has been allocated to meat ingestion containing sodium nitrate.

- Propyl Gallate

Another chemical preservative that does serve a good purpose to the food it is added. Propyl Gallate a stabilizer which is used mainly in packaged meals, dry milk, baked goods and etc. The NTP has establish, that consumption of this preservative can lead to prostate inflammation and tumors in the brain, pancreas, and thyroid.

- Potassium Bromate

Potassium Bromate is considered as carcinogenic and it leads to tumors in the kidneys and thyroid. The United Kingdom and Canada have prohibited potassium bromate to be used in bread to strengthen its dough, but this preservative is still used in United states to support the bread's dough.

- BHT and BHA

Butylated hydroxyanisole (BHA) and butylated hydroxytoluene (BHT), these two food preservatives are used to stop oxygenation.

Both BHT and BHA keeps fats from turning rancid and thus prolongs the shelf life of any food containing fat. The harmful effects of BHT are still unproven, but BHA has been displayed to directly cause cancer in animals.

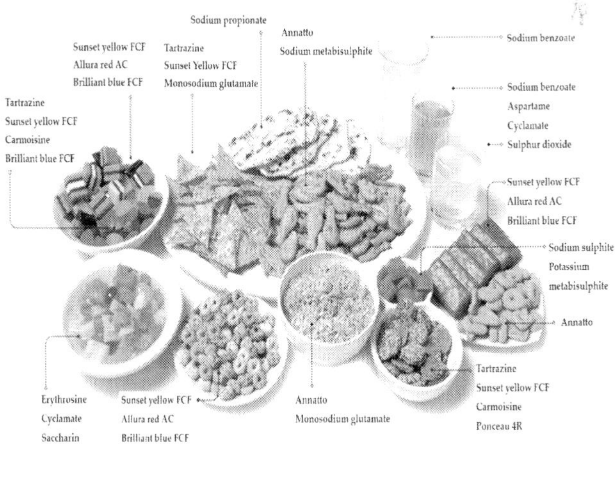

Figure 2.18: Food additives that should be avoided

Source: http://www.hungryforchange.tv/article/top-10-food-additives-to-avoid

CLASSIFICATION OF FOOD AND FEED ADDITIVES

Food Additives

Food additives as talked previously, are the substances that are added to the food to enhance the quality, stability, and shelf-life of food. Different class of additives used in food with their functions and examples-

- Anti-caking agents- like bentonite, calcium silicate and calcium aluminium silicate, are used to keep the powdered products, for example, flow freely when poured.
- Anti-foaming agents- like polyethylene glycol 8000 and triethyl citrate, they are used to reduce foaming in foods.
- Color retention agents- like ferrous gluconate etc., are used to maintain or increase the color of food.
- Bleaching agents- like chlorine, are used to whiten the foodstuffs.
- Firming agents- like calcium chloride, calcium sulfate, and calcium gluconate, are used to prevent the food from collapsing during processing and also to strengthen the food structure.
- Artificial sweeteners- like sorbitol, alitame, aspartame, and calcium saccharin, are used to impart sweet taste to the food stuffs for fewer kilojoules or calories than sugar.

A complete class of additives are listed below in the table-

Table 2.3: Classes of additives

Class of additive	Function	Examples
Anti-caking agents	Keep powdered products (e.g. salt) flowing freely when poured	• Bentonite (558), • Calcium aluminium silicate (556), • Calcium silicate (552)
Anti-foaming agents	Reduce or prevent foaming in foods	• Polyethylene glycol 8000 (1521), • Triethyl citrate (1505)

Antioxidants	Retard or prevent the oxidative deterioration of foods	• Butylated hydroxyanisole (320), • Ascorbyl palmitate (304), • Calcium ascorbate (302)
Artificial sweeteners	Impart a sweet taste for fewer kilojoules/ calories than sugar	• Sorbitol (420), • Alitame (956), • Aspartame (951), • Saccharin / calcium saccharin (954)
Bleaching agents	Whiten foods	• Chlorine (925), • Chlorine dioxide (926), • Benzoyl peroxide (928)
Bulking agents	Increasing the bulk of a food without affecting its nutritional value	• Ammonium chloride (510), • Isomalt (953), • Polydextrose (1200)
Colorings	Add or restore color to foods	• Curcumin (110), • Brilliant blue FCF (133), • Tartrazine (102)
Color retention agents	Retain or intensify the color of a food	• Ferrous gluconate (579)
Emulsifiers	Prevent oil and water mixtures separating into layers	• Lecithin (322), • Sorbitan monostearate (491), • Ammonium salts of phosphatidic acids (442)
Enzymes	Break down foods (e.g. ferment milk into cheese)	• α-amylase (1100), • Lipases (1104), • Proteases (papain, bromelain, ficin) (1101)
Firming agents	Strengthen the structure of the food and prevent its collapse during processing	• Calcium chloride (509), • Calcium gluconate (578), • Calciumesulfate (516)

Flavor enhancers	Improve the flavor and/or aroma of a food	• Calcium glutamate (623), • Disodium 5′-ribonucleotides (635), • Ethyl maltol (637)
Food acids	Maintain a constant level of sourness in a food	• Acetic acid (260), • Citric acid (330), • Fumaric acid (297)
Flour treatment agents	Improve flour performance in bread making	• Sodium metabisulphite (223), • Ammonium chloride (510), • Potassium bromate (924)
Glazing agents	Impart a shiny appearance or provide a protective coating to a food	• Beeswax, white and yellow (901), • Carnauba wax (903), • Shellac (904)
Gelling agents	Thicken and stabilize various foods (e.g. jellies, deserts and candies)	• Agar (406), • Calcium alginate (404), • Carrageenan (407)
Humectants	Prevent foods from drying out (e.g. dried fruits)	• Glycerin or glycerol (422), • Lactitol (966), • Oxidised polyethylene (914)
Mineral salts	Improve the texture of a food (e.g. processed meats)	• Cupric sulfate (519)
Preservatives	Protect against deterioration caused by microorganisms	• Sodium nitrate (251), • Benzoic acid (210), • Sodium benzoate (211)
Propellants	Gases which help propel a food from a container	• Carbon dioxide (290), • Nitrogen (941), • Nitrous oxide (942)

Sequestrants	Bind and remove unwanted minerals that cause oxidation	• Potassium gluconate (577)
Stabilizers	Maintain the uniform dispersion of substances in a food	• Xanthum gum (415), • Guar gum (412), • Bleached starch (1403)
Thickeners	Improve texture and maintain uniform consistency	• Tannins (181), • Sodium alginate (401), • Pectins (440)
Vitamins	Restore vitamins lost in processing and storage	• B vitamins, including niacin • Vitamin C • Vitamin E

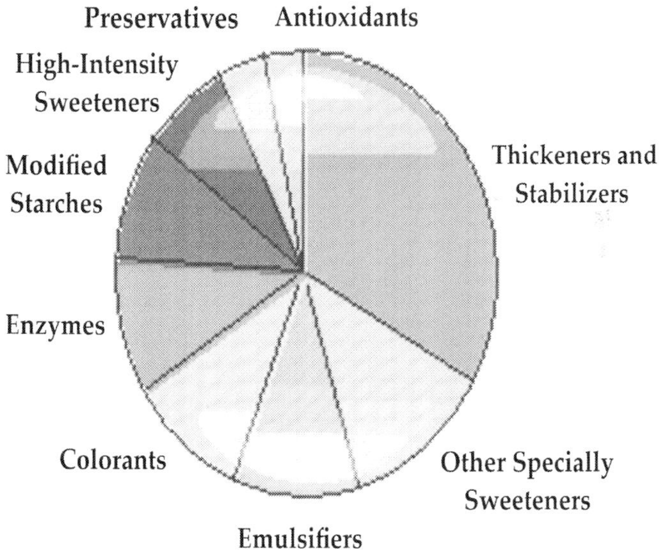

Figure 2.19: Consumption of food additives by types

Source: https://www.ihs.com/products/chemical-food-additives-scup.html

Feed Additives

Feed additives are the small amount of materials which are added in the production process of feed, the quantity in feed is actually very rare but it is delivered with noteworthy effects, feed additives are essentially used in the present feed industry so as to fortify the basic feed nutrition value, it has different effects for the enhancement of animal's performance, assuring the animal health, protect feed charges, increase quality of animal products, etc.

Feed additives are broadly classified into five main categories:

- Technological additives

Technological additives are the group of additives which effects the technical features of the feed. These additives don't directly affect the nutritional value of the feed but they may enhance its handling or hygiene characteristics, indirectly. An example of additive that indirectly affects the nutritional value and used for preservation of feed is - organic acid.

- Sensory additives

These are the group of additives that increases the palatability of a diet by inspiring the appetite, this is possible usually by the effect these additives have on the flavor and color of the diet. For instance, vanilla extracts.

- Nutritional additives

Nutritional additives supply detailed nutrients that are essential for the optimal growth of animals, for example- amino acids, vitamins, or trace minerals. In most of the cases, nutrients additives are basically the concentrated form of nutrients supplied as natural ingredient in the diet.

- Zootechnical additives

Zootechnical additives main purpose is to expand the nutrient status of the animal, by facilitating more proficient use of nutrients present in the diet and not just by providing specific nutrients. For example, an enzyme or direct fed microbial product, equally improve the conditions of the intestinal tract, and as a result it facilitates other effective nutrient extraction from the diet. Due to this, they are repeatedly mentioned to as pro-nutrients, i.e. products which recovers the nutritional value of a diet without essentially providing nutrients openly.

- Coccidiostats and Histomonostats

These products are used as a regulator in intestinal health of poultry through direct effects on the parasitic organism concerned, they are not categorized as antibiotics.

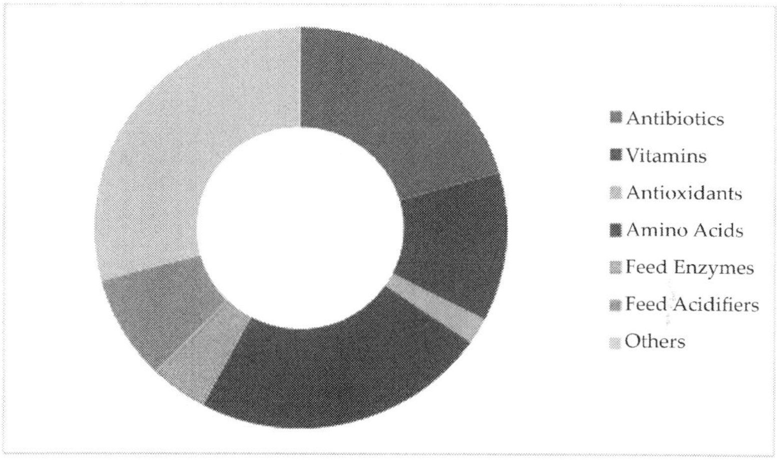

Figure 2.20: Poultry feed additive market share, by type

Source: http://www.marketsandmarkets.com/Market-Reports/poultry-feed-market-147838333.html

CHAPTER 3

BIO-PRESERVATION OF FOOD AND FEED ADDITIVES

"Looking good and feeling good go hand in hand. If you have a healthy lifestyle, your diet and nutrition are set, and you're working out, you're going to feel good."

- Jason Statham

INTRODUCTION TO PRESERVATIVES

Many of the food products are fragile by nature and need protection from spoilage during their preparation, storage, and distribution, to give desired shelf-life to the food. Keeping in mind the modern era, it is necessary to extend the shelf-life of the various food commodities, because food products today are often sold and distributed in areas that are far distant from the site where they are manufactured. This is done to prevent the food from early spoilage. Food preservation is a continuous process that is used to fight against microorganisms that spoils the food or make it unsafe. Some of the most common example of classical preservation agents are the weak organic acids, like- acetic, benzoic, lactic and sorbic acid. The acids stop the outgrowth of both fungal and bacterial cells. In the food production, it is vital that proper measures are taken so as to confirm the stability and safety of the product during its whole shelf-life.

In particular, modern consumer trends and food legislation have made the successful accomplishment of this objective much more of a challenge to the food industry.

Preservation is the method which is used to preserve the existing condition of the food or to prevent the food from various factors that causes damages, such as from chemicals (oxidation), physical (temperature, light) or biological (microorganisms). Preservation is mainly concerned to delay the spoilage of foodstuffs and to avoid any modification in its taste or appearance. This can be done by various methods, like- canning, smoking, dehydration (drying), and freezing, or by using food additives like- antioxidants or other preservatives. The principal aim of using preservatives is to make the food, by removing the effect of biological influences. The biggest risk to consumers is food being spoiled, or food becoming toxic by the effect of micro-organisms (e.g. bacteria, yeast, moulds). Some of these organisms can secrete noxioustoxic compounds, which are unsafe to human health and can even be lethal.

Preservative are the substances that are added to the food to save the food from the effect of harmful chemicals changes, for example- oxidation or growth of mold. Along with stabilizing agents and emulsifying agents, preservatives are used to sustain the freshness and consistency of food. There are various types of preservatives that are suitable for different products and work effectively against certain specific chemical change. Some of the few preservatives are- Antimicotics, these preservatives inhibit the growth of molds in products like- fruit juice bread cheese, and dried fruits. Examples of anti microtics are- sodium propionate, calcium propionate, and sorbic acid. Antioxidants, like- butylated hydroxytoluene, or BHT, to protect the food containing fats and oils from getting rancid. Antibiotics, like- tetracyclines are used to avert the growth of damaging bacteria in fish, poultry, and canned foods.

Preservatives not only helps to preventing spoilage, but also some preservatives play an aesthetic role, which means to upgrade the appearance of the product, example- sodium nitrate. To prevent the foodstuffs against spoilage by microorganisms, antimicrobials are used, antimicrobials are the substances that are used to delay, stop or prevent the growth and multiplication of bacteria, molds, and yeasts. Compounds like sulfites (Sulfphur compounds) are used to obstruct the growth of bacteria for example in wine, dried fruits, vegetables in brine or vinegar.

Another acid, named as sorbic acid is used for many different purposes, which includes preservation of potato products, jam and cheese. Other than sulfphur compounds, nitratenitrate, and nitrite compounds also another important group of substances that are used for food preservation. Nitrates and nitrites are used as additives in meat products like sausage and ham to shield the bacterium that causes botulism, this is how it contributes to food safety. Another group of compounds, such as benzoic acids and their calcium, potassium or sodium salts are used antifungals and anti-bacterialsbacterial in foods like- pickled cucumber, low sugar jams and dressings, jellies, condiments.

Figure 3.1: An apple without (left) and with (right) preservative

Source: http://thegreenserviette.blogspot.in/2015/07/food-additives-preservatives.html

To certify that preservatives, make the food safer and healthy, they are subjected to pre-market safety valuation and authorization practice. In Europe, the bodies accountable for the safety valuation, authorization, regulation and labelling of preservatives comes under European Food Safety Authority (EFSA) and the European Commission, Parliament and Council. The international level is governed by a Joint Expert Committee, from the Food and Agriculture Organization (FAO) and the World Health Organization (WHO), on Food Additives (JECFA). The preservatives safety valuation is generally based on reviews of all toxicological data, including remarks in humans and animal copies. From the data that is obtained, additive with no noticeable toxic effect is determined. This phenomenon of determination of additive with least or no toxic effect is termed as "no-observes-adverse-effect-level" (NOAEL) and it is used to determine "Acceptable Daily Intake" (ADI) character for each additive in the food. The ADI delivers a great boundary of safety and denotes to the quantity of a food additive that can be taken effectively daily in the diet, without causing any harm to human health. The authorization of usage of preservatives is administered by Directive 95/2/EC of the European Parliament and the Council of 20 February 1995 on Food Additives other than Colors and Sweeteners. There are many food additives that causes adverse reactions to the public health, although vigilant research and surveys shows that it is mostly the misconception that people have for few additives rather than any recognizable adverse reactions that additives cause. Preservatives rarely cause any true allergic or immunological reactions. Some of the food additives testified to cause adverse reactions are the preservatives that belong to the cluster of sulphating agents, the sulfating agents includes several inorganic sulfphite additives (E220-228), and benzoic acid and its derivatives (E210-213), may activate asthma characterized by breathing problems, shortness of breath, wheezing and coughing in asthmatic individuals.

Figure 3.2: Food preservatives

Source: http://www.foodadditivesworld.com/gifs/food-preservatives.jpg

The European Parliament and the European Council together has laid down a complete labelling scheme for additives to permit consumers to make knowledgeable choices with respect to foods that contains preservatives. Legislation also instructs that food additives that are added to food must be labelled on the packaging of food products by their category (preservative, colourcolor, antioxidant, etc) with their E-number or by their name. Some of the most widely used preservatives in the European Union are given below-

Table 3.1: E-number of different food substances and their usage

E-Number	Substance/class	Foodstuffs where they are used
E 200-203	Sorbic acid, sorbate compounds	Cheese, fruit sauces, toppings, wines, dried fruit.
E 210-213	Benzoic acid, benzoate	Pickled vegetables, candied fruits, semi-preserved fish products, sauces, low sugar jams and jellies.
E 220-228	Sulfur dioxide, sulfite compounds	Wine, dried fruits, fruit preserves, potato products.
E 235	Natamycin	Surface treatment of sausage and cheese.
E 249-252	Nitrate, nitrite compounds	Sausage, bacon, pickled herring, ham, foie grass, cheese.

FOOD PRESERVATIVES

Almost all the food products we use have preservatives in them except the ones that are grown in the kitchen, these preservatives increase the shelf-life of the food products apart from that they also protect the food from the influence of harmful microorganisms. All the manufacturers add preservatives during processing to the food. The chief purpose of preservatives is to avoid the spoilage of food stuffs during the transportation time.

Food is essential for the survival of human and animal body, keeping this in mind food preservatives are been used in food since historical times and they are one of the oldest technologies used by human beings to avoid its spoilage. There are several ways and means that have been originated and enhanced for this purpose. Boiling, freezing & refrigeration, pasteurizing, dehydrating, pickling is some of the traditional methods that are still used in some food stuffs for their preservation. Sugars and salts are also frequently used as food preservatives in many foodstuffs. Nowadays, nuclear radiation is another way of food preservation which is use for several foodstuffs. Modified packaging methods such as- vacuum packing and hypobaric packing also work as food preservatives.

Food Preservation is principally done for three motives-

- To preserve the natural characteristics of food
- To preserve the appearance of food
- To upgrade the shelf worth of food for storage.

Food preservatives are broadly divided under four categories-

- Natural Food Preservatives

The class of natural food preservatives includes the salt, sugar, alcohol, vinegar etc. These are the most common traditional preservatives that are used in food made at home, like- pickles, jams and juices etc. Some of the most common methods like- freezing, boiling, smoking, salting is measured to be the natural ways of preserving food. Coffee powder and soup are dehydrated and freeze-dried for their preservation purpose.

In this segment, the citrus food preservatives such as- citrus acid and ascorbic acid work on enzymes and disturb their absorption leading to preservation. Sugars and salts are the most primitive forms of natural food preservatives that drops the growth of bacteria in food very proficiently. For instance, salt is still used as a natural food preservative, for preservation of meat and fish.

Figure 3.3: Some of the natural food preservatives

Source: http://www.stylecraze.com/articles/harmful-food-preservatives-and-their-side-effects/#gref

Chemical Food Preservative

Another form of preservatives that are in use for quite a long time are chemical food preservatives. Chemical food preservatives are the best form of preservative and are very effective for elevating the shelf life of food and are largely full proof for the preservation purpose.

Some of the examples of chemical food preservatives are:

- Benzoates (like sodium benzoate, benzoic acid)
- Nitrites (like sodium nitrite)
- Sulphites (like sulphur dioxide)
- Sorbates (like sodium sorbate, potassium sorbate)

Antioxidants are also the chemical preservatives that are used in food, and they act as free radical scavengers. In this group of food preservatives comes the vitamin C, BHA (butylated hydroxy anisole), bacterial growth inhibitors such as sodium nitrite, sulphur dioxide and benzoic acid.

Ethanol is also used as one of the chemical food preservatives in food, wine and food stored in brandy. Like natural food preservatives not all the chemical food preservatives are beneficial, some of them are even injurious to health. For example- Sulfur dioxide and nitrites are harmful for human health the chemical food preservatives are that are harmful. Sulfur dioxide causes irritation in bronchial tubes and nitrites are carcinogenic.

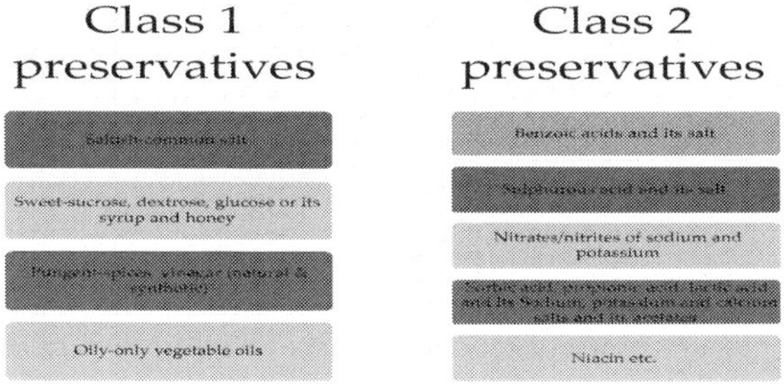

Figure 3.4: The class 1 and 2 preservatives

Source: http://thegreenserviette.blogspot.in/2015/07/food-additives-preservatives.html

Artificial Preservatives

Artificial preservatives are the chemical constituents that halts the overdue growth of bacteria spoilage and its the food discoloration. The artificial preservatives are either added to the food or sprayed on the food. Types of artificial food preservatives are:

- Antimicrobial agents: includes- Benzoates, sodium benzoate, sorbate and nitrites.
- Antioxidants: includes- Vitamin E, vitamin C, sulfites and BHT (butylated hydroxytoluene).
- Chelating agent: includes- Disodium ethylene diamine tetra acetic acid (EDTA), citric acid and polyphosphates.
- Harmful Food Preservatives

Although preservatives are the food additives that are used to keep the food fresh and to stop the bacterial growth. But still Although, there are various preservatives in food that are damaging the human health if taken in more than the prescribed parameters. Some of the examples pf for harmful food preservatives are:

- Benzoates- Benzoates are banned in Russia to be used in foods as a chemical food preservative. Use of benzoates in foods triggers allergies, skin rashes and asthma. They are also measured as a cause of brain damage. Benzoates are commonly used in foods like- tea, coffee, fruit juices etc. Two types of benzoates are used for preservation of food, that is- potassium benzoate and sodium benzoate.

  Potassium Benzoate inhibits the growth of yeast, mould and various bacterial growth. Potassium benzoate is the potassium salt of benzoic acid. The E-number of potassium benzoate is E212.

  Sodium Benzoate is bacteriostatic and fungistatic in acidic conditions. It is also known as "Benzoate of Soda". The E-number of sodium benzoate is E211.
- Butylates- Butylates are expected to cause high blood pressure and increased cholesterol level in blood. Use of butylates in foods affects the kidney and liver functioning. Butylates are found in foods like- butter, margarine and vegetable oils.

- Butylated Hydroxy anisole- BHA is predicted to cause the cancer and liver diseases. BHA are used for the preservation of fresh pork and pork sausages, instant tea, potato chips, cake mixes and much more.

  Both BHA and its related compound named butylated hydroxytoluene (BHT) are the phenolic compounds that are added to the food to preserve the food from fats and oils and also to keep they away from becoming rancid. Both phenolic compounds are used in food, cosmetics, and in packaging of those products with fats to maintain color, flavor, and nutritional level. Molecular formula of BHA is C11H16O2 and BHT is C15H24O.

  Both BHA and BHT are antioxidants. Oxygen reacts favorably with BHA or BHT and hence protects the food from spoilage. BHA and BHT are also used in cosmetic and pharmaceutical industry to preserve fats and oils. The main role of BHA is used keep from away from becoming rancid, also it is used as a yeast de-foaming agent. BHA is found in butter, meats, cereals, chewing gum, baked goods, snack foods, dehydrated potatoes, and beer. Some of the other sources where BHA are found are animal feed, food packaging, cosmetics, rubber products, and petroleum products.
- Caramel- Caramel is used as a coloring agent in food preservation. Use of caramel in food causes the vitamin B6 deficiencies, genetic effects and cancer. Caramel are found in foodstuffs such as bread, candies, brown colored food and frozen pizza.

Caramel coloring is found in many foods, it is the most widely used food coloring worldwide. Caramel is made by the heating process called caramelization. The color of caramel ranges from pale yellow to amber to dark brown in its basic state. Once added to foods, it creates various other colors.

In addition to the above mentioned harmful food preservatives there are many other preservatives too that cause damage to health. Some of these are- bromates, carrageenan, caffeine, coal tar AZO dyes, chlorines, gallates, glutamates, mono-glycerides and di-glycerides, nitrates and nitrites, saccharin, sodium erythrobate, tannins and sulphites. Some common preservatives and their primary activity are given in the below table:

Table 3.2: Preservatives and their primary activity

S. No.	Chemical affected	Organism(s)	Action	Use in foods
1.	Sulfites	Insects and microorganisms	Antioxidant	Dried foods, wine and juices
2.	Sodium nitrites	Clostridia	Antimicrobial	Cured meats
3.	Propionic acid	Molds	Antimicrobial	Bread, cakes and cheese
4.	Sorbic acid	Molds	Antimicrobial	Cheese, cakes and salad dressings
5.	Benzoic acid	Yeast and molds	Antimicrobial	Soft drinks, ketch ups and salad dressings

Role of Food Preservatives-

- Food preservatives deactivates the role the pathogenic micro-organisms found in fresh vegetables and raw meats, lessens food-borne diseasesdiseases, and makes food safer to consume.
- Processed foods are less vulnerable to spoilage than fresh foods, modern processing, storage and transportation distributes extensive variety of food from around the world, giving so many choices in the superstores that would have been unthinkable to our ancestors.
- Processing often intensifies the taste of food, though in some cases it can also have the reverse effect.
- The nutritional value of food can be improved by adding extra nutrients and vitamins during processing.
- The nutritional value can also be made more constant and dependable.

- Modern processing technologies can also upgrade the quality of life for people who have allergies, by eliminating the proteins that root allergic reactions.
- The mass production of food makes the processed food much cheaper to produce at the cost of making meals from raw ingredients at home.

In modern food preservative industry, there are at least twenty-five commonly used food preservatives that are designed to prevent the food from spoiling. Out of those twenty-five, there are three mostly used categories- Mould inhibitors, these mould inhibitors diminishes the growth of mould on bakery products and also used in cheese and dairy products to avert the mould growth. Some of the examples of mould inhibitors are- calcium propionate, sodium propionate and sodium benzoate. Salts that are in use from the ancient times in food, salts are used to draw out all the water from the meat so as to destroy the bacteria living in the meat. The third most important preservatives are the acidifiers, acidifiers are used to circumvent spoilage as the microbes cannot withstand in acidic environment. Some of the example of acidifiers used as preservatives are- vinegar, wine and citric acid.

PRESERVATIVES IN FEED AND THEIR SUPPLEMENTS

Feed preservatives are the products that are used for animal nutrition for improving the quality of feed from animal origin and enhancing the quality of feed. Unless the scientific evaluation is properly done assuring that the additives don't have harmful effects on human and animal health or on environment and proper authorization is given to them feed additives are not put in the market. some of the feed preservatives used in feeds are discussed in detailed below:

Ethoxyquin-

Ethoxyquin was established over thirty-five years ago. Ethoxyquin was developed as a rubber stabilizer by Monsanto. It was later approved as both fungicide and a preservative, ethoxyquin is used in feeds to guard fats from getting rancid or lipid peroxidation. The basis of ethoxyquin is quinoline, which is made from coal tar. Other than quinoline there are other preservatives that are made from coal tar like- vitamin B found in pet supplements, equine supplements and human supplements. Quinoline is also the basis for the herbicide called Assert, and is widely used as a growth regulator to safeguard appears and apples in harvest and storage.

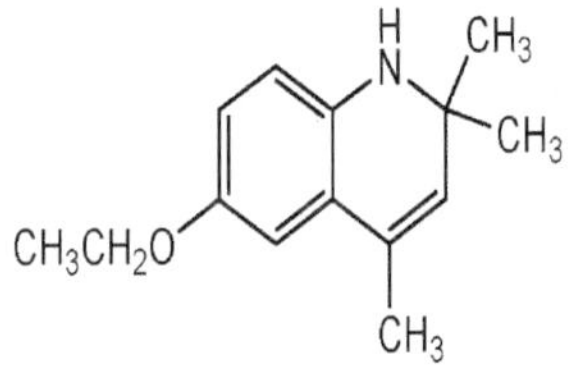

Figure 3.5: Chemical structure of ethoxyquin

Source: https://www.chemservice.com/ethoxyquin-n-11877-100mg.html

Propionic Acid-

Propionic acid is designed synthetically from carbon monoxide and ethyl alcohol. Propionic acid is usually classified by the EPA as a bactericide and fungicide. The world's largest producer of propionic acid is BASF, it produced 149,000 metric tons in 2010. Propionic acid is a key constituent in apple cider vinegar (along with lactic, acetic and citric acids) and is also a chemical constituent of human sweat. Propionic acid used in feeds and hay is the synthetic form of propionic acid, i.e. not obtained from apple cider vinegar. According to the New Jersey Department of Health, propionic acid has the Hazardous Substance Fact Sheet, and the greatest dangers to humans are caused by using it on spray crops.

Propionic acid is registered in the list of carcinogenic chemicals. As propionic acid is used in many portions of the country on feed, it seems to be comparatively safe for equine ingestion.

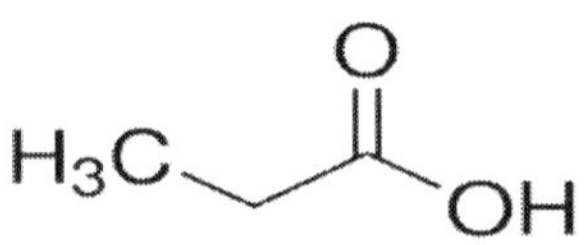

Figure 3.6: Chemical structure of propionic acid

Source: http://www.sigmaaldrich.com/catalog/product/sial/402907? lang=en& region=IN

Citric acid-

Citric acid is a chief constituent of citrus fruits, but this preservative cannot be extracted from oranges, lemons or limes. Citric acid is made through a fermentation process of molasses, corn-steep liquor, and hydrolysedhydrolyzed corn starch. As the fermentation process ends, citric acid is isolated by precipitating it with calcium hydroxide to produce calcium citrate salt that is then redeveloped with sulphuric acid. The corn by-products like- corn-steep liquor, and hydrolysedhydrolyzed corn starch used are from genetically modified corn. By the end of year 2007, worldwide production of citric acid was 1,600,000 tons being china the largest single-country producer, manufacturing over 50% of that total citric acid produced.

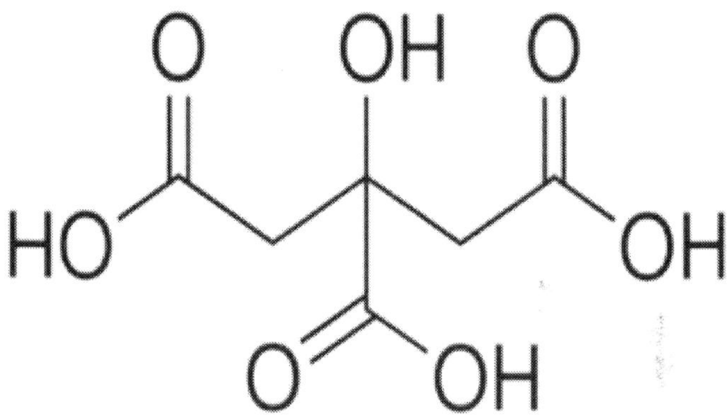

Figure 3.7: Chemical structure of citric acid

Source: http://everlastingchemicals.com/products/chemicals/4

Natural vitamin E-

Natural Vitamin E is also called as semi-synthetic vitamin E or d-alpha tocopherol. The oil form is obtained from soy, palm or vegetable oils, which is then esterified using acetate. Acetate in turn are made from acetic acid that can be formed by bacterial fermentation but is predominately made by methanol carboxylation. This method, uses methanol and carbon monoxide to produce acetic acid. Methanol is resultant of natural gas, though it used to be generally produced from coal. The elementary esterification process is a chemical reaction where two reactants, most commonly an alcohol and an acid, form an ester. Although, synthetic vitamin E are not generally used as a preservative, but some companies use these synthetic forms, like- dl-alpha tocopherol or dl-alpha tocopheryl, are used by some companies in the feed and supplement industry for vitamin E supplementation. The principal oils that are extracted for vitamin E content i.e. corn, soy and palm, are extracted using hexane. Corn and soy oils are obtained from genetically modified seeds unless stated organic. Palm oils, today are the most important source of edible oil or biofuel, and have become very popular worldwide, but recently palm oils come under severe criticism as the deforestation of Indonesia, which is the largest producer of palm oil, goes unchecked. Natural vitamin E are not identical to food sourced vitamin E. food sourced vitamin E are whole food itself, like- wheat germ, wheat germ oil, almonds, sunflower seeds, and camelina oil, etc.

Acetic acid-

It is normally used as a preservative in hay. Approximately, seventy-five percent of the acetic acid is formed by carbonylation using methanol and carbon monoxide. Acetic acid is one of the major component of vinegar, but this biological source accounts only for ten percent (10%) of the world production. There is no known indication of toxicity in animals eating hay with acetic acid.

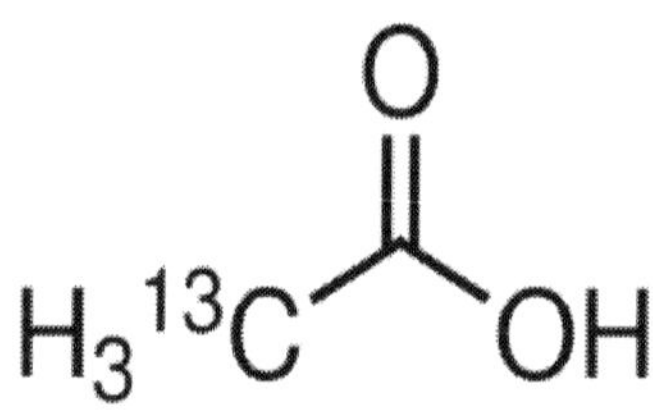

Figure 3.8: Chemical structure of acetic acid

Source: http://www.sigmaaldrich.com/catalog/product/aldrich/279307?lang=en&region=IN

BHA and BHT-

Both food preservatives are banned in almost all countries including Canada, England, Japan, and Australia. The National Institutes of Health's National Toxicology Program decided that: "BHA can be reasonably anticipated to be a human carcinogen." The European Commission has positioned BHA in Category 1 potential endocrine disruptor based on signs that it restricts the hormone function in at least one living organism. Some studies suggest that BHT also acts as a weak endocrine disruptor and may affect lungs, liver, kidney and thyroid. Some data have revealed that it leads to cancer in rats, also it results in developmental effects and thyroid changes.

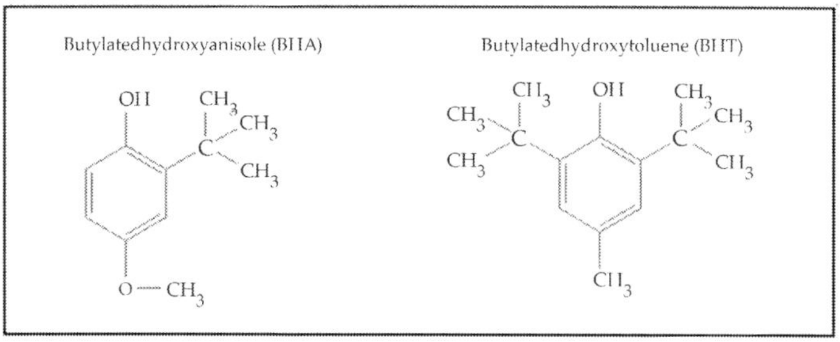

Figure 3.8: Chemical structure of BHA and BHT

Source: http://omg-chemistryyyy.blogspot.com/2013/03/whats-on-your-face.html

MOST EFFECTIVE BIO-PRESERVATIVES: NISIN AND PROPIONIC ACID

1) Nisin

Nisin was discovered first in 1928, in England, as an outcome of complications that are experienced during the cheese making process. Milk storage had allowed the impure or contaminated microorganisms to grow and produce the inhibitor. The protein that was isolated was named as "nisin" from its sero-group determined by the Lancefield sero-typing arrangement for streptococci. Nisin is a group N inhibitory material. It was in 1951 when the potential application of nisin in food preservation was demonstrated. Alpin and Barrett have commercialized the nisin concentrate as "nisaplin", which is used as a preservative in cured meats, canned foods, milk and dairy products, and other sections of fermentation industry.

Nisin is produced by the bacterium Llactococcus lactis which is a polycyclic antibacterial peptide used as a food preservative. Nisin contains 34 residues of amino acid, it also includes some uncommon residues also other than the common ones, the uncommon residues are lanthionine, methyllanthionine, didehydroalanine, and didehydroamino butyric acid. These uncommon or not so usual amino acids are familiarisedfamiliarized by posttranslational modification of the precursor peptide. In the above reactions, a ribosomally synthesisedsynthesized 57-mer is transformed into a final peptide. The unsaturated amino acids are usually originated from serine and threonine, and by the enzyme catalysedcatalyzed addition of cysteine residues to didehydro amino acids resulting in multiple thioether bridges. Subtilin and epidermin are interrelated with nisin. All the members belonging to this class are known as lantibiotics. In the food industry, nisin is generally not chemically synthesised, it is obtained from the culturing of Lactobacillus lactis on natural substrates, like- milk or dextrose.

Lactococci that produces nisin occurs naturally in raw milk and cheese. According to the worldwide survey of raw milk samples that were reported 43% of them contained Lactococci. Nisin was initially isolated in the late 1930s, and since 1950s they are produced as Nisaplin from the naturally occurring sources, like- alpin and barret in the laboratories. Nisin was later in 1960s, was approved as a food additive for use in the United States of America. Nisin is an infrequent example of "broad-spectrum" bacteriocin that are effective against various gram-positive microorganisms and against spores, while other bacteriocins generally hinders only closely related species. For instance, some of the gram-positive microorganisms against which nisin is effective includes-lactic acid bacteria, that are frequently associated with spoilage, Listeria monocytogenes, which is a known pathogen, Staphylococcus aureus, Clostridium botulinum, Bacillus cereus, etc. Gram-negative bacteria are habitually protected by their outer membrane but they become susceptible to nisin action one heat shock is given or when they are couple with a chelator i.e. EDTA. Nisin are soluble in water and are active at points approaching the parts per billion range. The concentration of nisin is measured by variety of techniques, some of which are- chromatography or by simple agar diffusion bioassay.

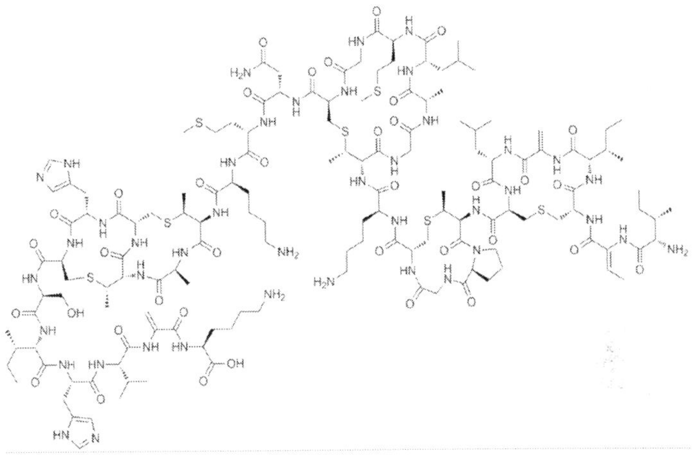

Figure 3.9: Chemical structure of nisin

Source: https://www.researchgate.net/figure/274641303_fig3_Fig-3-Chemical-structure-of-nisin

Nisin is mostly used in processed meats, chesses, and beverages, etc. during production. The objective of adding nisin is to extend shelf life by overpowering Gram-positive spoilage and pathogenic bacteria. The shared range of nisin that is accepted to be used in foods without harming the food at all is ~1-25 ppm, that depends on the type of food. However, the toxicological studies conducted newly have confirmed that nisin is not toxic, even at levels much higher than those used in foods. Nisin has an E number of 234 as a food additive.

Nisin is certified as 'generally recognized as safe' (GRAS) by the US Food and Drug Administration, and is therefore executed by more than 48 countries all over the world in many food applications. Nisin is non-toxic even with its antibacterial characteristics and the proteinaceous nature of nisin is fully degraded in the human intestine.

Nisin is a natural food preservative that can kill cancer cells- Nisin, which is a naturally happening food preservative growing on dairy products, transports a punch to two most deadly diseases: cancer and deadly, antibiotic-resistant bacteria. Scientists were focused on the properties of nisin and therefore they discovered that nisin has cancer-killing properties. Nisin is a colorless, tasteless powder which is extensively used as a preservative in food. Some of the discoveries and sayings about nisin by several researchers are listed below-

"The application of nisin has advanced beyond its role as a food biopreservative" said researcher Yvonne Kapila, professor at University of Michigan School of Dentistry in the US.

"Current findings and other published data support nisin's potential use to treat antibiotic resistant infections, periodontal disease and cancer," Kapila noted.

Figure 3.10: Nisin is a naturally occurring food preservative that grows on dairy products. A new study revealed this 'magic mold' can wipe out cancer cells and antibiotic-resistant bacteria

Source: http://www.dailymail.co.uk/health/article-3395898/Magic-MOLD-wipe-two-lethal-conditions-Natural-preservative-grows-dairy-products-kills-cancer-cells-antibiotic-resistant-bacteria.html

Scientists found that feeding rats a "nisin milkshake" can kill upto 70-80% of neck and head tumour cells after nine weeks and prolonged survival. Nisin is typically added to the food at the rate of .25 to 37.5 mg/kg. some food products contain nisin, but nowhere near the 800 mg/kg which is required to kill cancer cells. There are numerous food products which are reachable to consumers also contain nisin - creams and pharmaceuticals to fight against infection and mastitis, and a sanitizer in lactating cows. Nisin also fights against fatal bacteria's like antibiotic-resistant MRSA (methicillin-resistant Staphylococcus aureus). MRSA is resistant to almost all the antibiotics, except for vancomycin for which it shows sensitivity.

2) Propionic Acid

Propionic acid guard's food and animal feed from mouldmold. The EU hygiene directive in force since January 2006 also appeals to animal feed, demanding it to be in perfect condition. Animal feed products must be free from mouldmold and have an extended storage life. Propionic acid is the perfect animal feed preservative. It is also a significant element in the making of vitamin E. In response to increasing call for propionic acid, BASF started up a new plant at the commercial site in Nanjing, China in summer 2005. The firm is now one of the world's foremost creators of this chemical intermediate. As everyone knows, no one can't count on the weather conditions. Its impulses may hit farmers very hardly. Too dry is evil, too wet is harsh. For instance, farmers don't appreciate rain at the time of harvest. The weather doesn't constantly play lengthwise, though. Farmers who desires to safeguard peace of mind have two substitutes: either to practice superior hot air systems to dry their grain preceding to storage, or treat it with the preservatives. The foremost purpose is to prevent mould from emerging. The toxins contained in mould would finish a farmer's whole feed reserves. An eco-efficiency investigation conducted by BASF likened the two methods. The outcomes show that use of propionic acid is more cost-effective and beneficial for the environment.

Farmers can pick from a collection of propionic-based product mixes. The accurate product for the application depends on the moisture content of the grain and the deliberate storage period. Wet grains and longer storage times demand for extra active substance. "Depending on the moisture content and storage period, farmers need six to 15 kilograms of propionic acid to preserve one metric ton of grain," says Ulrich Roser, BASF's technical advisor for feed preservatives. The preservative-treated grain is steady for about one year and is used by farmers for their own feed mixes, arranged by adding soy meal, mineralsminerals, and vitamins. The mixes are mostly used for pig feed.

Propionic acid keeps the feed fresh or new for longer period. Propionic acid, like formic acid and acetic acid, fits to the family of carbonic acids. All three are powerful fungicides and can be calculated on to kill mould. Nevertheless, propionic acid is a much more influential agent than its shorter-chained relatives. The grain needs to be wetted all over with propionic acid to avert mouldsmolds from evolving. Mould toxins in crops leads to fertility disorders, digestive problems and immune deficiency in animals. Propionic acid also successfully fights other pests organisms such as yeasts, bacteria, and insects like the granary weevil, thus defending both animal feed and farm animals.

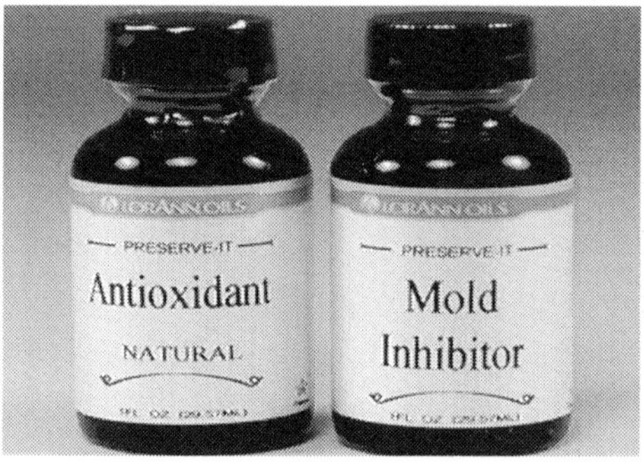

Figure 3.11: Bottles representing antioxidant and mold inhibitor

Source: http://www.pastrychef.com/FOOD-PRESERVATIVES_p_2087.html

Propionic acid itself is completely digested by the farm animals. In fact, dairy cows produce this acid in their own rumen as the chief by-product of natural digestion: bacteria in the bovine stomach releases the two acids, named, propionic acid and acetic acid from the cellulose in the grass they eat. A single cow produces approximately one and a half litresliters of the acid per day and benefits from it as a source of energy. One kilogram of propionic acid has the nutritional value of two kg of barley.

In the development process that goes to produce Swiss cheese, bacteria break lactic acid products into propionic acid and carbon dioxide. The acid stops the cheese from going bad during the extended maturation period. Because of the acid taste, the volume of acid in the cheese must not surpass a certain boundary. The acid concentration may increase to as much as one weight percent – a level satisfactory to preserve grain with a moisture content of 26 percent for the entire year.

Its trustworthiness as a feed preservative also qualifies propionic acid for other agricultural applications, for instance, it is used as an additive in liquid animal feed, hay and drinking water. Microorganisms and algae are no longer able to grow and water not gets contaminated. Propionic acid is tremendously effective in fighting with mouldsmolds and pests. It is also a key element in herbicide manufacturing.

According to the EU hygiene directive, the farmers and feed mix manufacturers must supply and use feed in perfect condition. Propionic acid-based feed preservatives are vital to accomplish this purpose. "Farm-mixed feed is a vigorously growing market in Germany. Our shares in this segment are growing at a huge rate year on year," says Roland Preuss, Product Manager Propionic Acid.

With the successful yearly production of 110,000 metric tons of propionic acid, BASF is worldwide market leader. More than forty percent of the acid formed is used in animal feed preservation. The chief markets for propionic acid are the Europe and USA. The Asian market is significantly smaller as compared to other markets but it is growing at a faster rate. The core Asian markets - China, India and Taiwan – have received their materials from Nanjing since summer 2005. "It has greatly reduced the times we need to supply our Chinese customer base, which includes explains animal feed manufacturers like Huamei, Kangtei and Youxianbao," says Anthony Jones, Manager Asia Pacific Acids Marketing. And Peter Zhang, Marketing Manager Acids in China, backs this up: "Our delivery service has improved enormously since production started up in Nanjing. Our customers are delighted with the speed and reliability of their deliveries."

Propionic acid is equally significant in Germany. "Propionic acid is the mainstay of grain preservation throughout Germany, now as in the past. And we expect acid preservation to take on even greater importance in the future," says Dr. Sabine Rahn, Product Manager Preservatives and Feed Specialties with AGRAVIS Raiffeisen AG.

Propionic acid in food is very beneficial, not just as an element in vitamin E production, but also in the form of a calcium bond, it benefits keeping the sliced bread fresh. Nearly one-fifth of the propionic acid sold is used in food preservation, mainly bread. The USA is the principal market for this. This multitalented acid is also in great call as a raw material for the chemical and pharmaceutical manufacturing. For many other requests, the acid needs no conversion. It is used in its unaffected state as a solvent, for example in large-scale technical schemes to produce aromatic aldehydes and acids. In gratitude of its numerous benefits, propionic acid's choice of applications is growing. Its chief function is in animal feed and food preservation and remains the same in long run. The use of propionic acid is set to rise in agriculture, not least because of a current ecoefficiency examination. The outcomes obtained have left no room for doubt: A farmer who thinks ecologically and acts economically will choose only propionic acid for preservation.

Another class of preservatives are the parabens which are used in cosmetic and pharmaceutical industry. Chemically parabens are a series of parahydroxybenzoates. They are very effective preservatives and are operative in many formulas. Paraben compounds and their salts are used primarily for bacterial and fungicidal properties. Parabens are mainly found in shampoos, shaving gels, commercial moisturizers, personal lubricants, topical pharmaceuticals, makeup, spray tanning solution, and toothpaste. Parabens are also used as an additive in food.

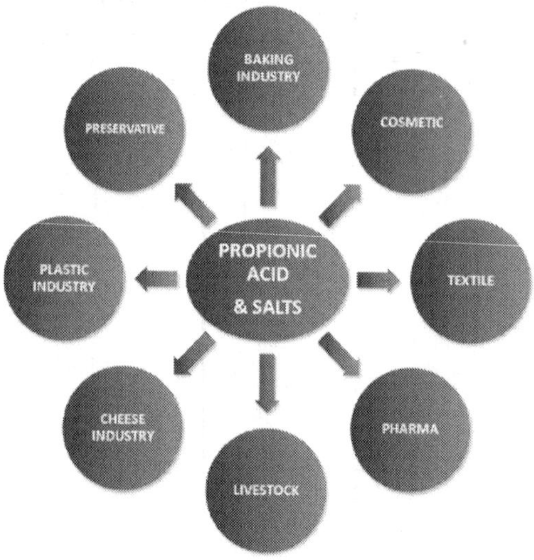

Figure 3.12: Various uses of propionic acid

Source: http://www.prathista.com/product/propionic-acid-and-salts

HARMFUL PRESERVATIVES IN FOOD AND FEED

Propyl Gallate:

Propyl gallate is commonly used in meat products, pickles, sauces and chewing gums. It is an injurious food preservative that can cause difficulties at birth and lead to liver damage in the extended period. Propyl gallate also lead to numerous other complications including breathing disorders and is therefore disqualified in several countries

SulphitesSulfites:

SulphitesSulfites are typically used to preserve dry fruits, canned fruits, corn syrup, pepper, canned olives, wine and vinegar. These compounds lead to diverse side-effects, for instance- palpitations, allergic reactions like asthma and rhinitis. SulphitesSulfites also persuade joint pain, headache, and cause cancer in few cases.

Brominated Oils:

Brominated oils are used in preservation of bottled juices for prolonged periods. Brominated juice is used quite extensively and can lead to possible destruction and serious side effects. Brominated oils can generate fluctuations in heart tissue, kidney damage, thyroid swelling, upsurge fatty deposits in your liver, and weakened testicles. Several countries have claimed that brominated oils should be banned for preserving bottled drinks.

Propylene Glycol and Carboxymethylcellulose:

Both the preservatives are consistent additives in ice cream. Propylene glycol is generally used as an antifreezing agent and a paint remover. It is the more threatening chemical out of the two preservatives, and various countries have obsoleted its use. The second preservative that is mentioned here is carboxymethylcellulose, which is a stabilizer. It is typically used in salad dressing, cheese spread, and chocolate milk. An investigation on rats concluded that carboxymethylcellulose initiates tumor production. CNS depression is one of the main side-effects of propylene glycol.

Mono-Glycerides and Di-Glycerides:

Both chemicals are generally used to preserve foods such as cakes, pies, cookies, bread and peanut butter. Mono and Di-Glycerides also supports preservation of other foods like roasted nuts, veggies packed with sauce and even margarine. Monoglycerides repeatedly feature on the list of evil or morbid fats.

Sodium Nitrate:

Sodium nitrate is another injurious food preservative that is used to preserve meat. Sodium Nitrate is added in many forms of processed meat like bacon, ham, and sausage. This nitrous acid is pretty unsafe for the body and often leads to stomach cancer. European nations like Norway and Germany have prohibited sodium nitrate to be used as a food preservative.

Maleic Hydrazide:

Maleic hydrazide preservative is generally added to potatoes to save them from sprouting. It is one of the well-known chemical inhibitor and can also lead to cancer.

Bromates:

Bromates are the most regularly used preservative in white flour and bread. It can persuade diarrhoea and destroys nutrients.

Citric Acid:

Citric acid is a vital nutrient found abundantly in most of the citrus foods. Though, citric acid formed using sulfuric acid is rather harmful and often leads to allergic reactions. This product makes citric acid hazardous if one is prone to allergies.

Benzoates:

Benzoates are the preservatives that are usually added to pickles, margarine, fruit puree and juices. Benzoates often induce allergies and lead to brain damage. These food preservatives elevate the shelf-life of food in grocery stores, but the effects of this preservatives on human health differs.

Ethoxyquin:

Ethoxyquin is a synthetic antioxidant used principally in animal feed, like- aquaculture and pet food. Ethoxyquin is not accepted globally to be used as a direct food additive for human consumption; consequently, it should not be measurable in the food supply. Some krill meals and crude fish oils are preserved using ethoxyquin, in the omega-3 industry. Because of the controversial use of ethoxyquin, some omega-3 manufacturers have questioned why it is used at all in sources that can source both human and animal nutrition products. Ethoxyquin successfully helps to decrease the risk of combustion in products like krill or fish that are not kept frozen. The supply chains for fish and krillskrill for human consumption tend to be well separated from each other, and the animal feed product economics do not defend the same types of refrigeration or other methods that are used for human consumption products, so chemical antioxidants are converted as the only option.

CHAPTER 4

SWEETENERS AS ADDITIVES

"*When you start eating food without labels, you no longer need to count calories*"

-Amanda Kraft

INTRODUCTION TO SWEETENERS

Sweeteners are food additives that are utilisedutilized in demand to inspire a sense of sweetness to the person consuming the product. Numerous chemical compounds are measured to be sweet, however only a fraction of these compounds is in use. Some sweet substances like Lead (II) Acetate are sweet in taste but then are highly toxic. Sweeteners are generally categorized into two foremost divisions, nutritive sweeteners and non-nutritive sweeteners. Nutritive sweeteners are those sweeteners that are assimilated to some level in the body and as such have a food energy value. Non-nutritive sweeteners are the ones that are not digested by the body and thus have an insignificant food energy value.

A fondness for sweet taste is inborn and sweeteners can upsurge the desire of eating. Nutritive sweeteners contain carbohydrates and deliver energy. They befall naturally in food or might be added in processing of food or by consumers before consumption. Higher consumption of added sugars is allied with advanced energy intake and inferior diet quality, which in turn increases the danger for obesity, prediabetes, type 2 diabetes, and cardiovascular disease. On an average, adults in the United States consume uptoup to 14.6 percent of energy from added sugars. Polyols or sugar alcohols, they add sweetness with lower energy and may lessen the danger for dental caries. The body does not distinguish amongst naturally occurring sugars and sugars that are added to food, but those added to food are most frequently linked with little nutrient-dense food. Consumers should set a boundary to these bare sources of energy to help attain or continue a healthy weight. Consumers who need a sweet taste but without added energy can pick from five FDA-approved non-nutritive sweeteners based on their private taste preference and the planned use. Non-nutritive sweeteners, when used as a replacement for nutritive sweeteners, might help consumers limit carbohydrate and energy consumption as an approach to achieve blood glucose or weight.

NUTRITIVE AND NON-NUTRITIVE SWEETENERS

Nutritive sweeteners

Nutritive sweeteners are those sweeteners that are assimilated to some level in the body and as such have a food energy value. Nutritive sweeteners like- table sugar, molasses, and polyols provide energy and comprises of 2-4 calories per gram of weight.

Sugar and Molasses

Sugar is an organic compound in the form of carbohydrate. Carbohydrates are comprised of molecules of carbon, hydrogen, and oxygen. There are diverse kinds of sugar, like- simple sugars or monosaccharides which contains glucose (dextrose), fructose (fruit sugar) and galactose, and complex sugars or disaccharides which contains sucrose (table/granulated sugar).

Sugar has a very elongated past in preservation of food- it has been used broadly in the preservation of fruits like apples and pears and in the making of a varied range of products like jams and jellies, fruit juice and sweetened foodstuffs such as condensed milk. When sugar is added to foods it binds with the water in the foods which results in falling in the volume of water that is available for the growth of microorganisms. One of the most common way of using sugar is to desiccate fruits by drying them and then packing with pure sugar. Fruits that are conventionally preserved through this technique include ginger, cherries, and the peel of citrus fruits. Alternatively, foods may be kept in a sugar syrup or cooked in sugar till they crystalize.

To produce jams and marmalade; fruits are primarily boiled to decrease the water content of fruit and to diminish contaminating microorganisms. Sugar is then added to the fruit to avert the regrowth of bacteria. Other food preservation approaches encompass merging sugar with alcohol to allow the preservation of fruit in alcoholic, like- Brandy.

Sugar is also used with salt to preserve various foods particularly fish and meat. Sugar may either be added to salt to produce a dry mixture which shelters food or may be liquefied form is used to make a brine to surround the food. Adding sugar to a salty brine supports to converse a sweetness to meat and fish and helps to decrease the punitive flavor of salt. While sugar can act as a preservative in a measured water constricted environment, for example in sealed jar, sugar itself attracts moisture so where water is accessible, sugar will attract water and the subsequent environment will become more attractive to microorganisms, like- yeast. The worth of sugar in food preservation is dependent on the quantity of sugar used, adding too little sugar to the food will still empower the growth and development of microorganisms. For instance, in the making of jams and jellies, adding too little sugar endorse the growth of mouldmold and yeast. Adding too much sugar to food affects the health welfares of the food, just like salt. Initially, sugar and sugary foods are a principal source of dental caries or tooth decay.

In addition, sugar also delivers empty calories- that is calories with no extra nutrients. An additional consumption of sugar is related with obesity and situations linked with obesity like- diabetes and heart disease. Due to the health effects of sugars it should not be used in very high quantity but one should keep in mind that if its level goes down the declared standards it makes the food more prone to microorganisms. In addition, while the use of an artificial sweetener to substitute sugar can help in maintaining the taste of a product, artificial sweeteners do not have the identical preservative action as of sugar, that is why sugar free jams and marmalade should also be necessarily refrigerated to aid their preservation.

Figure 4.1: Different types of sugar that are added to food

Source: http://www.stuff.co.nz/life-style/well-good/motivate-me/64858517/making-sense-of-sugars-and-sweeteners

Natural Sweeteners are commonly divided into two groups: with calories, and without calories. Some of the most common natural sweeteners are Honey, Stevia, Maple Syrup, Coconut Palm Sugar and Molasses. All the natural sweeteners mentioned above contain calories with the exclusion of Stevia. While selecting the finest natural sweetener one needs to keep a check on calories, effect on blood sugar, effect on dental hygiene, beneficial effects, taste, price, and usefulness in cooking. Some of the natural sweeteners used at home and for commercial purpose are:

Table sugar or sucrose is the characteristic white granulated product that is used to sweeten foods. Brown sugar is refined form of white sugar with molasses mixed in and, though the flavour varies somewhat, it has almost the same sweetness level as white sugar on basis of volume. Powdered sugar is excellently grounded sucrose joined with a minor quantity of corn starch to avoid caking. This starch makes it inappropriate for canning because it might cause the brine to become cloudy. Sugar serves as a protecting agent and helps in gelling of jams and jellies, in addition to its sweetening effect.

Cane Sugar comes from the sugar cane. A giant perennial grass with stout, jointed, fibrous stalks that are rich in sugar. They come from the same plant as white sugar, but are less refined compared to white sugar and contains more vitamins, minerals, and trace elements. This lets our insulin levels to not point as swiftly as white refined sugar. Each of the cane sugars below has a diverse grade of refinement.

Organic Whole Cane Sugar, Rapadura/Sucanat- This is the only category of cane sugar that is unrefined. Organic whole cane sugar is a deep brown colour with a caramel taste that is not crystalized and holds all its nutrients.

Rapadura is prepared by initial extraction of juice from the sugar cane, and later evaporating the water out of the juice by stirring with paddles over little heat. Conventionally, this juice was then molded into a block. The most widespread form of organic whole cane sugar in the United States is from the German corporation, Rapunzel, initially called Rapadura. Because of diplomatic glitches it was retitled as Organic Whole Cane Sugar still under the Rapunzel brand. Rapunzel uses a recent technique of drying the sugar by passing it through a sieve, leaving behind a grainy sugar. Rapadura is the one cane sugar that does not separate the sugars during molasses. This permits all its vitamins and minerals to be engaged letting our bodies to digest it more leisurely. It does not affect much of the blood sugar levels like refined sugars. Rapadura is a dark colour sugar and has a royal deep flavour. It palates outstanding in morning coffee and is also used in baking. Because of its deep flavor, it makes a great substitute to regular brown sugar in any of the baking needs. Sucanat, or Sugar Cane NATuralnatural, is an American emblem and following to Wholesome Sweeteners, it is made in the same way as Rapadura.

Figure 4.2: Different variety of cane sugar

Source: http://www.myhumblekitchen.com/2011/12/a-simple-introduction-to-natural-sweeteners-and-how-to-use-them/

Raw sugar is the residue that is left after sugar cane has been processed to eliminate the molasses and improve the sugar crystals. Raw sugar contains contaminants like moulds and fibres, although most of the raw sugar sold in the United States is in purified form.

Two popular categories are the coarse-textured dry Demerara Sugar and the utmost fine-textured Barbados sugar. Turbinado sugar, a raw sugar that has been steam-cleaned, has uneven blond ccolored crystals with a mild molasses flavourflavor.

Raw sugars are the refined sugars, which are natural brown, and large crystal sugars. In the United States, it is recognized as Turbinado, Raw Sugar. In the United Kingdom, it is typically identified as Demerara. They are handled in approximately the similar way as white sugar excluding they don't go through the whole process. Since some of the molasses is left over it does not contain few nutrients. It tastes awesome in coffee and tea, and because of the large granules present in these, they are great to be topped on baked muffins or pies.

Figure 4.3: The raw sugar

Source: http://www.wisegeek.com/what-is-raw-sugar.htm

Evaporated Cane Juice/Organic Sugar- Evaporated cane juice or organic sugar is also an example of refined sugar. Again, the evaporated cane juice also doesn't go through the complete refining process as some of the molasses is left in the ultimate product leaving about trace vitamins and minerals.

Evaporated cane juice tastes like white sugar and hence can be used as 1 on 1 substitute for white sugar.

Figure 4.4: The above picture shows the packaged organic cane sugar

Source: https://www.target.com/p/wholesome-organic-sugar-2lb/-/A-17440652

Honey is a natural product where the chief sugar is fructose. It is harmless to use honey as sweetener for canning or freezing. Nevertheless, the flavourflavor of the honey sweetened foods may be noticeably dissimilar than anticipated.

Raw honey is unheated, unfiltered, unpasteurized, and unprocessed form of sugar that is obtained from honey bee. Raw honey comprises of 27 minerals, 22 amino acids and 5,000 live enzymes.

Apart from the rich source of minerals and vitamins, raw honey fights against many disease, heals the body from within, soothes illness, boosts the digestive system and used as a natural beauty product, due to its beneficial qualities, it is considered as a super food.

When replacing honey in a baked recipe, substitute 1/2 cup honey per cup of sugar. So, if a recipe demands 1/2 cup sugar, use a 1/4 cup honey.

Figure 4.5: Raw honey

Source: http://www.myhumblekitchen.com/2011/12/a-simple-introduction-to-natural-sweeteners-and-how-to-use-them/

Corn syrup is a viscous blend of glucose, fructose, and polysaccharides. It is hard to alter a volume of sugar into a corresponding amount of corn syrup at the same sweetness level. Therefore, corn syrup should only be used for home food preservation if a recipe specifically indicates its use.

Agave syrup or agave nectar, is a processed sweetener that arises from a succulent plant. It covers almost the same caloric value as the table sugar does but is sweet in taste because of its very high fructose content. Therefore, one-fourth less of agave nectar than sugar is used. Scientific research has not yet been done on the usage of agave nectar in home canning. Thus, there is no permitted recommendation for its use.

Figure 4.6: The agave syrup

Source: https://www.sharecare.com/health/artificial-sweeteners/article/agave-nectar-health-benefit-or-hype

Real Maple Syrup- Real maple syrup from the sap of the sugar, black or red maple tree comprises of vitamins and minerals. Real maple syrup is high in manganese and zinc and its plain sweet flavour makes it a great substitute to brown sugar. When replacing maple syrup in a recipe, substitute 1/2 cup honey per cup of sugar. So, again if a recipe demands for 1/2 cup sugar, use a 1/4 cup real syrup, just like raw honey.

Figure 4.7: Real maple syrup

Source: http://www.myhumblekitchen.com/2011/12/a-simple-introduction-to-natural-sweeteners-and-how-to-use-them/

Dried fruits- dried fruits are also used to give a sweet taste to the food products. By adding dried fruit to baked goods, oatmeal, granola, and yogurt, sweetness can be added without using sugar or its constituents.

Figure 4.8: Benefits of organic coconut palm sugar

Source: http://www.eatingdisorderpro.com/2013/02/08/the-benefits-of-coconut-palm-sugar/

Coconut Palm Sugar- Conflicting to its title, coconut palm sugar does not have coconut taste. Coconut sugar is not made from the coconut, instead it is drawn from the sap of the coconut palm tree buds. It is rich in nutrients occupied with minerals, such as potassium, magnesium, zinc, iron, and vitamin B.

Polyols

Polyols are widely used in different marketable food applications and its use have been increasing the market ever since. They have many useful and organoleptic features. These features are the main and only reason for its high use in the market. Several types of polyols are easily accessible and hence allows for the expansion of products that too with the extensive variety of features and improved appearance of the product. Polyols can substitute all the features of sucrose in food products. Hence it helps the maker to stabilize the complete product with respect to its sensory features i.e. flavor, color and texture. Therefore, there exists a great use of these polyols specifically in the "no sugar added", "condensed sugar" and "sugar-free" products. They are even suitable for the aids like "reduced calorie" requirements and for the people who suffer from diabetes.

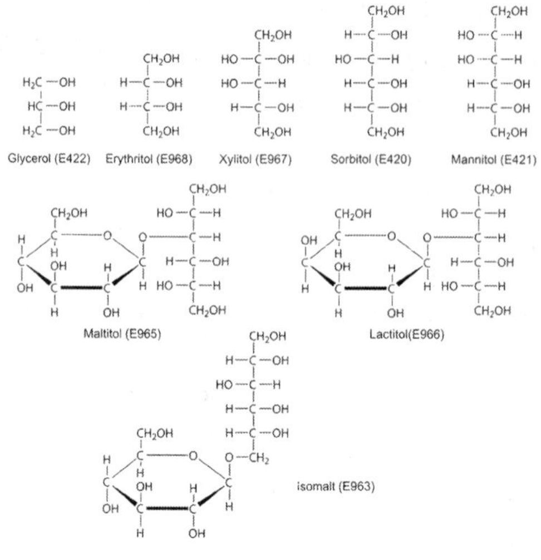

Figure 4.9: Polyols as food additives

Source: https://st2.depositphotos.com/1678626/7835/v/950/depositphotos_78351242- stock-illustration-polyols-as-food-additives.jpg

Non-nutritive sweeteners

All non-nutritive sweeteners are right for freezing. For both canning and freezing, precise recipes are used, like those available from the National Centre for Home Food Preservation to make definite the suitable sweetening agent that can be used for the envisioned purpose and for appropriate amount to add. Non-nutritive sweeteners can also be added before serving the food. Non-nutritive sweeteners can be added to fruit pulps before being dried to make fruit leathers.

The Food and Drug Administration (FDA) has accepted an innovative food additive as a sweetener and flavourflavor enhancer in foods named, Advantame. Advantames are used in food products that ncludes baked goods, non-alcoholic beverages, chewing gum, confections and frostings, frozen desserts, gelatins and puddings, jams and jellies, processed fruits and fruit juices, toppings, and syrups. Whether it's to cut down on the number of calories they consume or any of a variety of other reasons, some people use sugar substitutes or high-intensity sweeteners to sweeten and add flavourflavor to the foods such that to cut down the number of calories or may be due to some other reasons. There are many permitted sugar alternates already in the market from which to choose. Some of the most commonly used forms of non-nutritive sugars are-

Saccharin, was first revealed and used in 1879, before the present food additive approval course came into effect in 1958. It was generally sold under the brand name 'Sweet N Low.' Saccharin is an artificial, or non-nutritive sweetener that is used in the making of many foods and pharmaceutical products including: Baked goods, Jams, Chewing gum, Drinks, Tinned fruit, Medicines, and Toothpaste. Saccharin is usually 200-700 times sweeter than sucrose or table sugar, saccharin does not increase blood sugar levels and like all other non-nutritive sweeteners it has no calories. Nevertheless, it does have an unpleasant or metallic aftertaste, particularly at high concentrations. Saccharin is unstable when heated but it does not react with other food ingredients chemically, hence it is good for storage. The substance was original discovered in 1878 by a researcher Constantin Fahlberg, who was working on coal tar derivatives in the laboratory at the John Hopkins University in Baltimore. Use of the substance; saccharin became prevalent during the sugar scarcities of World War I. Its fame further increased in the 1960s and 1970s amongst dieters as an effect of its 'calorie-free' status.

Saccharin has many benefits, including that it can be merged with other sweeteners. Saccharin is frequently blended with other artificial sweeteners to recompense for each sweetener's weaknesses. For instance, it is generally mixed with cyclamate in countries where both of these sweeteners are legal, with each sweetener used to cover the other's off-taste. Mixtures of saccharin and aspartame are also frequently used in diet carbonated soft drinks to guarantee that some sweetness leftovers in the event that syrup is stored beyond aspartame's comparatively shorter shelf-life. Ingesting saccharin-sweetened products can be an advantage to diabetic's person as the substance goes straight through the human digestive system without being assimilated.

Saccharin has no food energy, but it can activate the release of insulin in human's due to its sweet taste. Although, saccharin is promoted and marketed as a 'calorie-free' sweetener, some recent studies have found that saccharin elevates blood glucose levels. It is assumed that these effects are due to variations in gut bacteria activated by the sweeteners. However, many of these studies are conducted on mice, and those that are conducted on humans includes very small sample sizes. Due to this, it is very tough to draw solid decisions from the study. Saccharin use in human food is plagued with various safety concerns. In the early 1970s, studies on rats in the laboratory allied saccharin with the expansion or development of bladder cancer in rodents. As a result, the U.S. Food and Drug Administration (FDA) pushed a ban on the use of saccharin in the United States based on the Delaney Clause of the Federal Food, Drug, and Cosmetic Act passed in 1958, which forbids the addition of any substances in the human food that persuade cancer in humans or animals. S saccharin was the only sweetener used during that time, there were dramatically very strong oppositions to its ban by public, the United States Congress interfered and permitted saccharin to continue in the food supply as long as all food comprising the artificial sweetener was labelled with a health warning. Since then, many human-based studies have concluded that the higher bladder cancer risk that is found in rats does not translate to humans. In 2000, the warning tags on saccharin-sweetened products were detached and in the subsequent year, the FDA inverted its position on saccharin, announcing it harmless for consumption. Saccharin is now one of five most widely used FDA-approved artificial sweetener, and is also an official additive in food in many countries including Europe and other countries around the world. Saccharin is heat stable and therefore it is used in jellies and jams. Saccharin is sold as Sweet 'N Low®, Sweet Twin®, and Sugar Twin®.

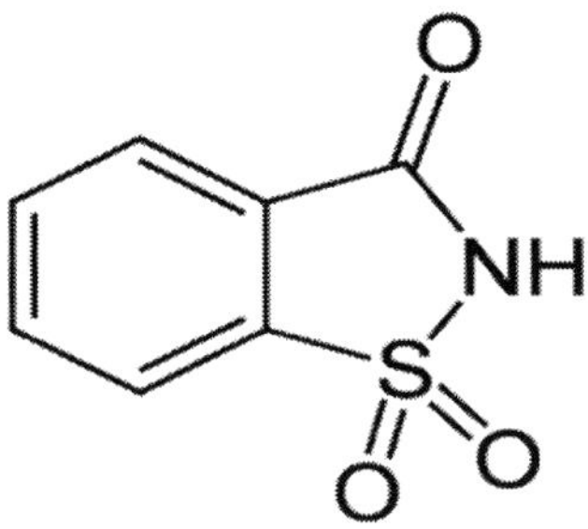

Figure 4.10: Chemical structure of saccharin

Source: http://www.mix1037fm.com/2016/10/31/the-more-you-know-what-is-sweetness/

Aspartame is an artificial, non-saccharide or non-nutritive sweetener and is used as a sugar substitute in few foods and beverages. Aspartame is a methyl ester of the aspartic acid or phenylalanine dipeptide. A board of experts set up by the European Food Safety Authority established in 2013 announced that aspartame is safe for human consumption at existing levels of exposure, as of 2017 sign of benefit was absent for weight loss with some data supporting weight gain and heart disease risks. Because its breakdown products of aspartame include phenylalanine, people with the genetic disorder phenylketonuria (PKU) must be attentive of this as an additional source. It was initially sold under the brand name NutraSweet. It was originally made in 1965, and the patent deceased in 1992. It got the initial consent for use in food products by the U.S. Food and Drug Administration (FDA) in 1981. The safety of aspartame has been the subjected to numerous political and medical controversies. Generally sold under the brand name 'Equal.' Aspartame is not recommended for canning since its sweetening properties are significantly condensed when heated. Aspartame is sold under the brand names Nutraweet® and Equal®.

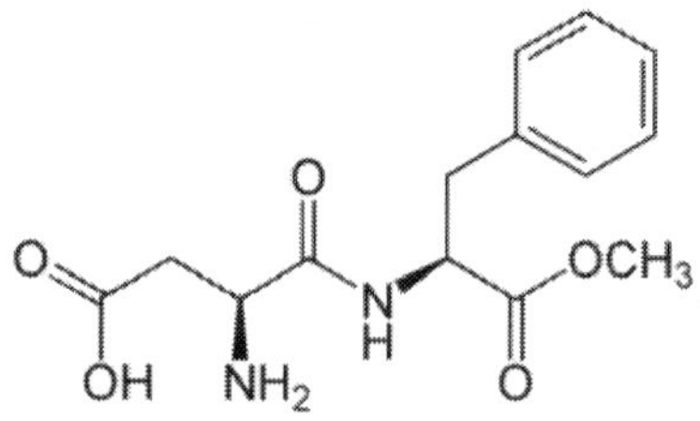

Figure 4.11: Chemical structure of aspartame

Source: http://www.mix1037fm.com/2016/10/31/the-more-you-know-what-is-sweetness/

Sucralose, was first accepted for use in 1998. Brand name is Splenda. Splenda® is a commercially expressed combination of sucralose, starch, and dextrose sugar. Though sucralose is a chemically improved form of sucrose with no nutritive value, the bulking agents added to it do contribute some energy value and the product covers around 12 percent of the calories of an equal volume of table sugar. Unlike other non-nutritive sugar alternatives, Splenda® is heat stable and is generally used in canned foods. Many people do notice an aftertaste flavourflavor that may upsurge with storage time. Although Splenda® delivers sweetness, it will not offer the firmness to canned fruits that sugar does. Products that are canned with Splenda® therefore are similar in consistency to those canned in water. The Splenda® website has recipes for making shelf stable jams and jellies.

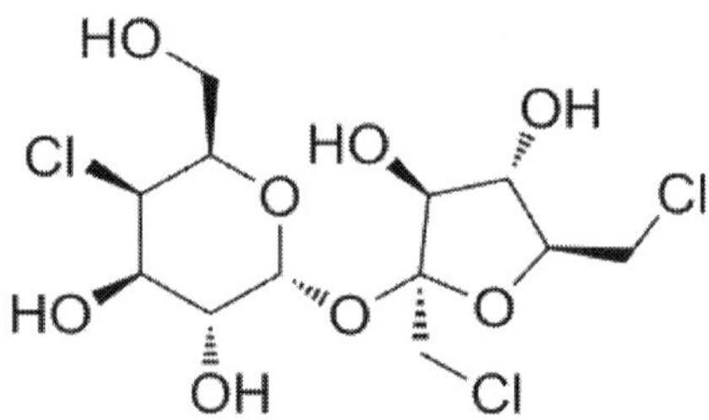

Figure 4.12: Chemical structure of sucralose

Source: http://www.mix1037fm.com/2016/10/31/the-more-you-know-what-is-sweetness/

Stevia is stable to heat and can be used for canning fruit and other foodstuffs where sugar is not serious to food security or consistency. Rebaudioside A, an active ingredient in Stevia, is 300 times sweeter than sugar. Stevia has been recorded as Generally Recognized as Safe (GRAS) by the Food and Drug Administration (FDA) and therefore is released from food additive guidelines. Truvia®, Sun Crystals®, A Sweet Leaf®, and PureVia® are Stevia-based sugar alternatives presently available in the market. According to the Truvia web site the product can be used for baking. There is no consistent information on their probability for canning and freezing. Green Stevia leaves or leaf powders are accessible but their sweetening effects might not be reliable.

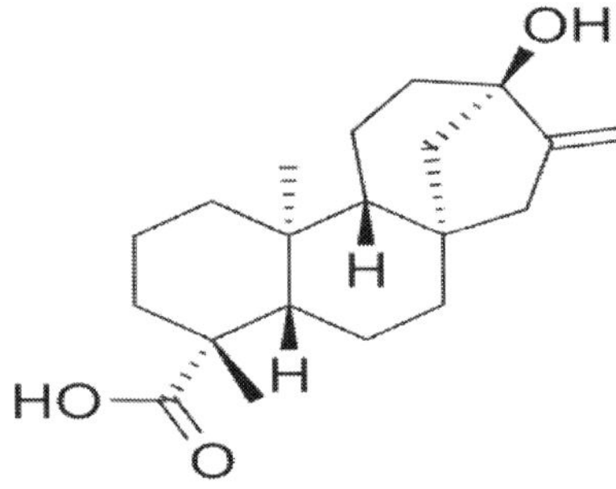

Figure 4.13: Chemical structure of stevia

Source: http://www.mix1037fm.com/2016/10/31/the-more-you-know-what-is-sweetness/

Acesulfame K or Acesulfame Potassium, was first permitted for use by the FDA as a harmless artificial sweetener in July'1988. Brand names in which it was sold includes 'Sweet One.' It is a derivative of acetoacetic acid. Unfortunately, there are numerous possible glitches linked with the use of acesulfame have been raised. They are based principally on animal studies since testing on humans remains restricted. The discoveries exposed the following: Acesulfame K arouses insulin secretion in a dose dependent fashion thus probably aggravating reactive hypoglycaemia. Acesulfame K seemingly formed lung tumors, breast tumors, rare types of tumors of other organs, some forms of leukaemia and chronic respiratory disease in some rodent studies, etc. According to the Centre for Science in the Public Interest, it was appealed on August 29' 1988 for a stay of agreement by the FDA because of "significant doubt" about its safety. Acesulfame K is the organic salt encompassing sulphursulfur and nitrogen and are 150-200 times sweeter than sugar. It is commonly used in beverages, baked goods, and candies and is used in the marketable food industry. Acesulfame K has a good shelf life and is comparatively the temperature and pH levels are stable. Acesulfame is sold as Sunett® and Sweet One®.

In addition to the five high-intensity sweeteners that are approved by FDA as food additives, the FDA permits use of two types of plant/fruit based high-intensity sweeteners: certain steviol glycosides obtained from the leaves of the stevia plant and extracts found from Siraitia grosvenorii Swingle fruit, also called as Luo Han Guo or monk fruit.

People who have phenylketonuria (PKU), an infrequent genetic disorder, have a problematic time digesting phenylalanine, a constituent of both aspartame and advantame. Newborns are confirmed for PKU using a common "heel-prick" test before they leave the hospital. Foods comprising aspartame must bear an information report for people with PKU, warning them about the occurrence of phenylalanine. But since advantame is sweeter than aspartame, so only a very little amount is needed to be used to reach the identical level of sweetness. As a result, foods comprising advantame do not need to bear that statement. While these high-intensity sweeteners are measured safe for their proposed uses, certain individuals may have a specific sensitivity or hostile reaction to any food substance.

The feed flavourflavors & sweeteners market has been rising at a stable rate from past few years. The consumer responsiveness in North America is fuelling the feed flavourflavors and sweeteners market in the area. The nasty taste in the feed is due to the additives added in the feed that are measured healthy for the gut of the digestive tract in the animals. Additionally, they also act as a decent alternative to the antibiotics in feed as the bitter taste is neutralisedneutralized by the flavourflavors and sweeteners. The market size for these additives is rising in evolving economies like- China, India, and Brazil.

The call for safe human food and mass making of meat has established significant consideration to the importance of feed additives. The promising probability of flavourflavors and sweeteners in swine and cattle feed delivers adequate indication to inspire the manufacturers to consider using such additives. The dietary supplement of these additives has an optimistic effect on the animal health, as they are mixed with other additives in feed with nutritional value that help in spreading of native lactic acid bacteria. These bacteria play a prominent role in the stimulating of the immune response against ailments. The feed flavourflavors & sweeteners market is anticipated to raise at a CAGR of 3.6% from 2014 to 2019. Asia-Pacific is progressively becoming a striking market for feed flavourflavors and sweeteners producers.

The cattle feed section controls the feed flavourflavors and sweeteners market owing largely to the high manufacturing of cattle meat in the world. The swine feed section is likely to raise at a CAGR from 2014 to 2019, owing principally to the high request of pork. Asia-Pacific lead the feed favours and sweeteners market, followed by North America.

North America is a foremost consumer of feed additives in the worldwide feed additives market, and holds a share of over 28 percent in 2011. It is the second largest market for feed flavourflavors and sweeteners with a share of about 31 percent in 2013, owing to the rising call for animal feed in the area. Therefore, the manufacturing of the animal feed flavourflavors in the area is fairly high. Additionally, the consumer awareness affecting to the chemicals used in animal feed has directed some of the chief innovations in the animal feed flavourflavors and sweeteners industry in the area. Nevertheless, the complete animal feed flavourflavors & sweeteners industry in the European region has been still for the past few years due to various reasons associated with the cost and productivity of the animal feed.

Some of the chief players in the worldwide feed flavourflavors and sweeteners market are- E.I. DuPont (US), Kerry Group (Ireland), Eli Lilly & Co. (US), etc. These companies deliver a variety of products, which offer specialised aids to animals counting swine, poultry, ruminant, and the aqua species, etc.

POLYOLS, INTENSE SUGARS, AND CARBOHYDRATES

Polyols

Polyols are widely used in different marketable food applications and its use have been increasing the market ever since. They have many useful and organoleptic features. These features are the main and only reason for its high use in the market. Several types of polyols are easily accessible and hence allows for the expansion of products that too with the extensive variety of features and improved appearance of the product. Polyols can substitute all the features of sucrose in food products. Hence it helps the maker to stabilize the complete product with respect to its sensory features i.e. flavourflavor, colourcolor and texture. Therefore, there exists a great use of these polyols specifically in the "no sugar added", "condensed sugar" and "sugar-free" products. They are even suitable for the aids like "reduced calorie" requirements and for the people who suffer from diabetes.

Polyols are used in variety of artefacts like in-

Confectionery- Due to the presence of several types of polyols it helps the maker to produce replacements for all the types of traditional food confectionery without even compromising on palate or superiority. The use of diverse range of polyols is so efficient that it permits its use to cater to necessities of the market like sweetness, flavourflavor and texture. Its use helps to produce fault free indulgent confectionary products which possess extra health benefits, which has led to a substantial growth in the confectionary market. Some of the examples of confectionaries in which polyols are used are- Hard Candy (High Boiling's), Chewing Gum, Coatings (Dragees), Soft & Chewy Candy, Gums, Pastilles and Jellies, Gelatine Gums, Gum Arabic Pastilles, Pectin Jellies, Tablets & Compressed Confectionery (including Mints), Lozenges, Aerated Confectionery, Fondant and Grained Confectionery, Toffee and Caramel, Fudge, Snack Bars.

Chocolate- the use of polyols has even transformed the chocolate industry too. It has changed the ways of development of chocolate. They have helped in the production of brilliant quality, no added sugar and sugar free chocolate. They have been widely used for the block chocolate, other than this polyol chocolates are also used for coating, enrobing and production of inclusions for bakery and ice cream applications.

Baked Goods Polyols- They are used in many bakery products to substitute sugar, and in most cases an appropriate polyol can be designated to permit the actual weight for weight replacement of sugar in these products. Final products have alike or better physiognomies to their traditional sugar covering its counterparts. Few examples of bakery products where polyols are used are- Cakes, Biscuits, Cereal and Nutrition Bars, Icings, Fillings, and Frostings.

Dairy & Ice Cream polyols- polyols helps in producing high quality and small calorie dairy products. It is likely to produce high quality, compact calorie dairy products using the polyols. The polyols propose a variety of freezing point depression factors, letting the development of frozen products with a wide spectrum of textures, melt and scooping appearances. Some examples are- Ice Cream, Frozen Desserts, Water Ices, Fermented Dairy Products, Jams, Syrups, Sauces and Fruit Preserves.

Not only this they are highly used in jams, syrups, sauces and food preparations. They tend to help a lot for the production of brilliant quality no added sugar and compact sugar products due to its sweetness and flavourflavor enhancing effects. Different level of groupings of polyols are used to deliver the anticipated sweetness and texture of the products.

Broad classification of polyols

Polyols are majorly divided into six types-Erythritol, Isomalt, Mannitol, Maltitol, Sorbitol, and Xylitol.

- Erythritol

Erythritol is a polyol or sugar alcohol that happens naturally, at low levels, in numerous fruits, mushrooms, and fermented foods like wine, soy sauce, and cheese. Erythritol is approximately 70 percent as sweet as sucrose, on the weight basis. It produces a cooling effect in the mouth, alike xylitol. Unlike other polyols, erythritol is voluntarily absorbed in the small intestine. It is not simply digested in the human body, so it delivers only 0.2 calories per gram. Erythritol is not hygroscopic i.e., it does not pick up moisture from the air. It has lesser water solubility than sucrose and a complex propensity to crystallize. Erythritol is principally absorbed in the small intestine and excreted out in the urine. It is not simply digested in the body. Since most of it is absorbed, it is not subjected to move for fermentation in the large intestine as most polyols are. This means that there is far less gas-forming and laxative effect, likened to the other polyols. Erythritol is occasionally used in grouping with a high-potency sweetener, to enable measuring a precise quantity of the high-potency sweetener, and to round out the taste outline.

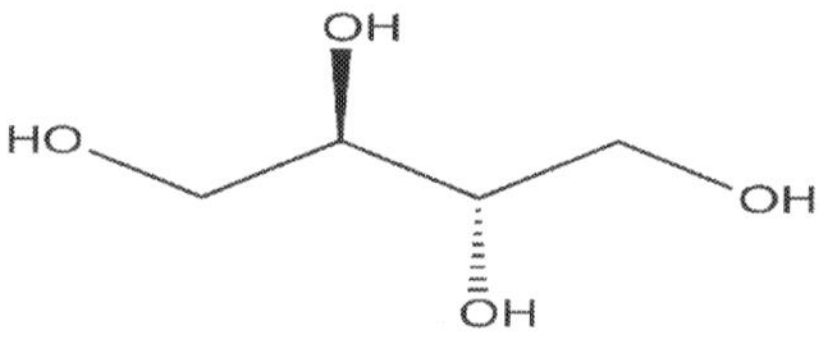

Figure 4.14: Chemical structure of erythritol

Source: http://www.sweetenerbook.com/erythritol_2.html

- Isomalt

Isomalt is a polyol or sugar alcohol. Isomalt is formed from sucrose in two steps: enzymatic rearrangement of sucrose to iso-maltulose, followed by hydrogenation of the product. It is a combination of two compounds, glucosyl-mannitol and glucosyl-sorbitol. Isomalt has a fresh sweet taste. It is about 40 percent as sweet as sucrose, on the weight basis. Isomalt is not proficiently engrossed by the body, and it is only partially absorbed. Its caloric value depends on numerous aspects. In the USA, isomalt delivers two calories per gram for labelling purposes. In the European Union, it is registered at 2.4 calories per gram. Isomalt's physical properties are identical like sucrose, making it suitable in hard candies and several other products. It is less hygroscopic i.e., less likely to pick up moisture from the air, than most polyols and sugars, so products made with isomalt are less sticky and have improved shelf-life. Isomalt has very decent heat constancy, so it is used in making hard candies and baked goods.

Isomalt is certified as GRAS in the United States since 1990. It has been in use in Europe from the early 1980s. The Joint FAO/WHO Expert Committee on Food Additives has resolute the "Laxative Threshold Value" (LTV) for a sum of polyols, and lactitol has an LTV of 29 grams per meal.

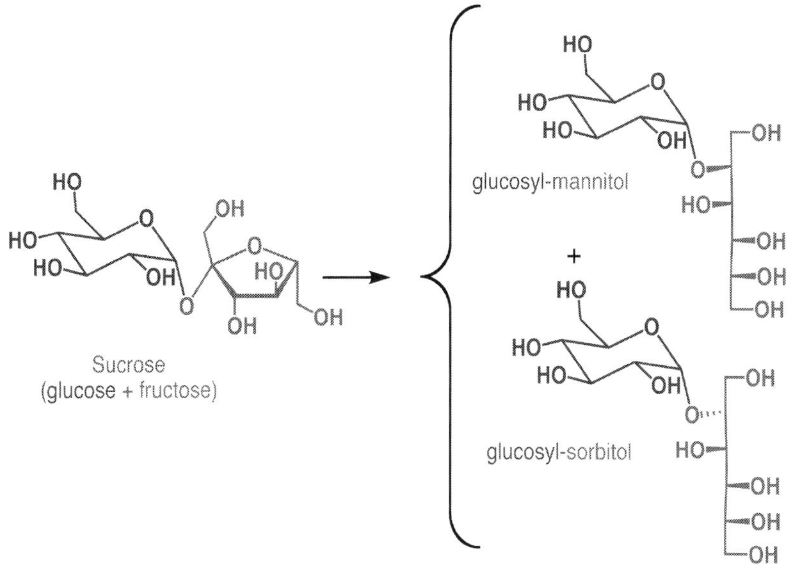

Figure 4.15: Chemical structure of isomalt

Source: http://www.sweetenerbook.com/isomalt_2.html

Isomalt are also commonly known as the sugar replacer, bulk sweetener and sugar-free sweetener. They are basically the sugar alcohol or hydrogenated carbohydrate just like other polyols. It is also the only polyol that is a derivative of beet sugar. It has a mild sugar like sweetness without any after-taste or disruption of the cooling effect. Nevertheless, if the cooling effect is required by the maker, flavourflavors and cooling agents can be added. It is same as sugar with context to its colourcolor i.e. it possesses white crystalline texture and substitutes sugar in the equal relative mass ratio. It usage is not only limited as a sweetening agent but it also enhances the texture and give a mouthfeel of sugar. Moreover, it is low-hygroscopic and hence can be easily stored. It's consumer products has a high overall shelf life as it does not absorb humid air or melt under extreme situations. Also, it doesn't encourage tooth decay as it is very little glycaemic and insulinaemic effect. "Linked to an energy value of 4 kcal/g for sugars and carbohydrates, in Europe isomalt is calculated with an energy value of 2.4 kcal/g, while in several other nations its scientific energy value of 2 kcal/g is used." It is used as a basic constituent in the varied series of high quality products such as candies, chewing gums, baked goods, chocolate, and many more other products. Isomalt has seen really high usage for the food applications like solid products or aqueous solution.

Candy: Isomalt is the most favoured ingredient in the world that is used for the sugar-free hard candies. The reason that the candy remains inn our mouth for the longer duration is the presence of isomalt in it. This happens because isomalt melts more leisurely than sugar. And hence the flavourflavor remains for longer duration in our mouth.

Candies which have isomalt as the ingredient are very steady and resilient to scratch and harm during manufacture and storage. Candies are not sticky even at the humid or high temperature due to the low hygroscopicity of isomalt. This makes their packaging even more easy as they can be kept in loose packaging like flip-top boxes and doesn't need to be enfolded separately.

Chewing gum: There exists several varieties of chewing gums such as pellets, stripes, balls, cushion-shaped, or direct-compressed. Isomalt is used in most of these types. Again, the due to the same reasons as mentioned above it melts leisurely and increases the flavourflavor release. Chewing gum gets uniform coating having a silky gloss and an outstanding crunch. It also shields the outer layer of the chewing gum from drying out early.

It also possesses some colourcolor-absorption properties that allow the coatings with strong colourcolors to retain their brilliance for a long period of time. It becomes resilient to abrasion also. It also helps for the softening of gum from the centrecenter and hence constancy can be maintained.

Bakery: Again, isomalt helps in maintaining the same taste, body, colourcolor, pore size distribution as sugar in the baked products. It also provides crunchiness and crispiness to the final products and hence provides the smooth surface. Moreover, it does not reduce the browning response of baking. So isomalt proves to be a very stable alternate.

Chocolate: There can be several types of chocolate like white, dark or milk; can be used as chocolate coatings, fillings like chocolate, caramel or nougat cream with nougat cream/spread. Isomalt is used in all of these types as substitute to sugar. Due to isomalt's soft and normal sweetness the chocolate flavourflavor completely develops without unwanted cooling effect.

- Mannitol

Mannitol is a polyol that is made by hydrogenation of fructose. It was originally isolated from the sap of a flowering ash tree which is found in southern Europe and Asia. Mannitol is nearby half as sweet as sucrose. Mannitol has a chilling effect in the mouth. Mannitol is not proficiently absorbed by the body, also just partially metabolized. Mannitol does not bind well with water. Mannitol is used as a dusting powder on chewing gums. It has a high heat constancy and can be used in chocolate flavoured coatings. Mannitol occurs certainly in mushrooms, algae, and trees.

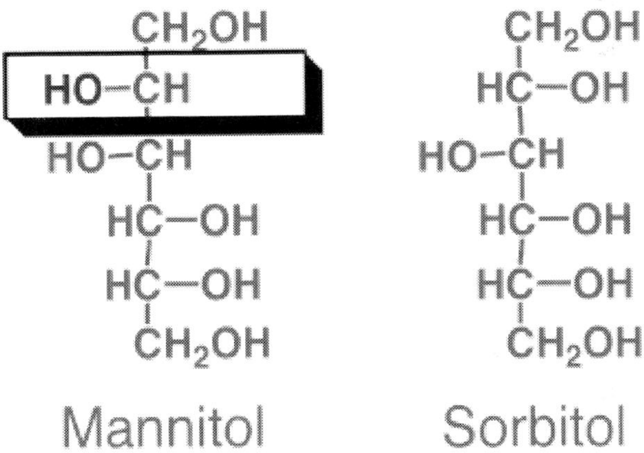

Figure 4.16: Chemical structure of mannitol (left) and sorbitol (right)

Source: http://www.sweetenerbook.com/mannitol_2.html

The FDA licenses use of mannitol in the USA. The Joint Food and Agriculture Organization/World Health Organization Expert Committee on Food Additives (JECFA) has resolved that mannitol is harmless. The Joint FAO/WHO Expert Committee on Food Additives has determined the "Laxative Threshold Value" (LTV) for many polyols. It evaluates the LTV for mannitol to be between 10 and 20 grams per meal. The FDA involves the following label statement for foods whose rationally foreseeable consumption may result in the daily absorption of 20 grams of mannitol: "Excess consumption may have a laxative effect."

Mannitol is a sugar alcohol or polyol, which has a good flavourflavor with slight or no after taste. It has 1.6 calories per gram, 40 percent the value of sugar but it has only 50 percent of the sweetness. Therefore, it is not actually valuable as part of a calorie measured diet, as it has 80 percent the calories of sugar for the identical sweetness.

However, mannitol has a very low glycemic index and thus it is very appropriate for diabetics. It is non-hygroscopic, meaning it does not absorb moisture from the atmosphere. This makes it very suitable as a hard coating for candy, and gives it certain desirable qualities in chocolate flavourflavor coatings. Like other sugar alcohols mannitol also has a laxative effect, and in fact is used for this purpose in some medicines, particularly for children. It is not suggested to consume more than 20 gm per day.

Though, mannitol is not used as a sweetener on its own, but it has substantial uses in the pharmaceutical industry as a drug, and as a covering for pills and tablets. It has very valuable applications in medicine and it is also used in the food industry.

- Maltitol

Maltitol is a polyol or sugar alcohol made by hydrogenation of the disaccharide maltose. Maltitol has a fresh sweet taste with negligible cooling effect in the mouth. It is about 70 percent as sweet as sucrose on a weight basis. Maltitol is only partly absorbed by the body, and it is only partially metabolized. Maltitol is not very hygroscopic, so it achieves well in chocolate and other foods with low-moisture content. Like the other polyols, maltitol does not contribute to browning responses.

Maltitol is certified as GRAS status in the USA. It is permitted for use in the European Union (EU). The Joint FAO/WHO Expert Committee on Food Additives has resolute the "Laxative Threshold Value" (LTV) for several polyols, and maltitol is one of the fewer laxative polyols with an LTV of 60 grams per gram.

Maltitol is formed from starch. Maltitol polyol sweeteners deliver the bulk, texture and preservative welfares of sucrose in sugar-free and calorie-reduced products. Maltitol is an appropriate sugar substitute for dental health, weight management and sugar avoiders. Maltitol syrup is a sweetener derivative of corn with all the taste and sweetness of sugar but with fewer calories.

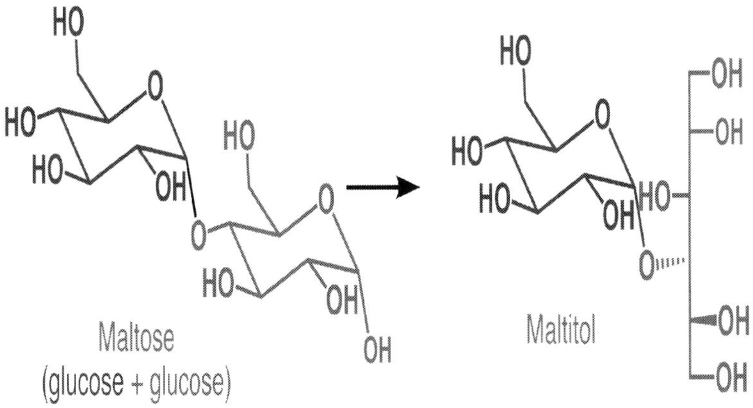

Figure 4.17: Chemical structure of maltitol (right)

Source: http://www.sweetenerbook.com/maltitol_2.html

- Sorbitol

Sorbitol is a polyol or sugar alcohol that occurs naturally in many fruits, including apples, pears, and plums. Sorbitol has a fresh, sweet taste. It is about 60 percent as sweet as sucrose on a weight basis. Sorbitol is only moderately absorbed by the body, and its absorption is equitably slow. Absorbed sorbitol is generally converted to glucose. Its caloric value depends on numerous factors, counting on how much is consumed, and individual difference in absorption rate.

In the United States, sorbitol delivers 2.6 calories per gram for labelling purposes. In the European Union, it is recorded at 2.4 calories per gram. Sorbitol acts much like sucrose in food classifications, with respect to providing bulk and interrelating with extra mechanisms to produce appropriate texture. It is predominantly good at binding moisture or humectant activity.

Sorbitol and the other polyols usually do not contribute to browning reactions that deliver typical colourcolor to baked goods.

The human body has the capability to absorb sorbitol and convert it to glucose, but it does not do so proficiently. Unabsorbed sorbitol can result in some digestive system unhappiness, like- gas, rumbling sounds (borborygmus), and diarrhoeadiarrhea. Sorbitol is a basic ingredient in sugar-free chewing gum.

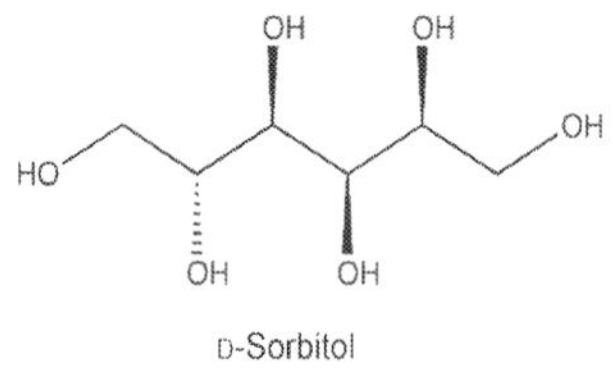

Figure 4.18: Chemical structure of sorbitol

Source: http://www.sweetenerbook.com/sorbitol_2.html

- Xylitol

Xylitol is a sugar alcohol or polyol. It is present naturally at low levels in raspberries, plums, and in some other fruits. It was primarily formed for food use in Finland from birch trees. It is now produced commercially from supplementary like corn and sugar cane stalks. Xylitol is around equivalent in sweetness to sucrose. It also produces a cooling effect in the mouth. It is used mostly in chewing gums and confections.

Xylitol is only partially absorbed by the body, and not transformed to energy very proficiently. Xylitol is steady over the range of pH found in foods. Xylitol is also heat-stable. It is fairly soluble in water i.e., 0.64 g per mL. Xylitol has been accepted by the US Food and Drug Administration, the Joint Expert Committee on Food Additives (JECFA), and the European Union. Unabsorbed xylitol leads to occurrence of some digestive system unhappiness, like- gas, rumbling sounds (borborygmus), and diarrhoea. Xylitol is lethal to dogs.

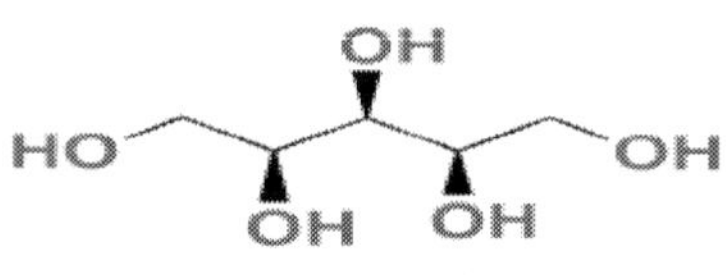

Figure 4.19: Chemical structure of xylitol

Source: http://www.sweetenerbook.com/xylitol_2.html

Intense Sweeteners

Intense sweeteners are the food additives that have a comparatively more sweetness many times than that of sugar which suggests they can be used in much smaller amounts. They are added to foods to substitute the sweetness usually provided by sugars without contributing meaningfully to obtainable energy. This means they can be observed as one way for consumers to control kilojoule or carbohydrate intake.

Many intense sweeteners are permitted for use in Australia, New Zealand and several other countries. These are alitame, acesulfame potassium (Ace K), aspartame, advantame, cyclamate, neotame, saccharin, sucralose, steviol glycosides and thaumatin.

- Aspartame

Aspartame is an artificial, non-saccharide sweetener that is used as a sugar alternate in various food products and beverages. According the European Union, aspartame is codified with the E-number- E951.

Aspartame is composed of two naturally occurring amino acids that is aspartic acid and phenylalanine, esterified with methanol. Aspartame has a sweet taste with negligible bitterness. The sweetness strength comparative to sucrose is around 180. It is composed of two amino acids which are the building blocks of proteins; like- proteins, it delivers around 4 calories per gram.

The high-potency, comparative to sucrose, means that it is not required to be used in higher quantity, so the calorie involvement in food systems is classically 0-2 calories. The constancy of aspartame in the dry state is very decent. In food systems, its stability is dependent on pH. It is highly stable at pH 4.3, in the pH range of numerous soft drinks. Its stability also depends upon the temperature, breaking down quicker at higher temperatures. This restricts aspartame's use in baked goods. Aspartame is completely digested in the body.

The safety of aspartame has been broadly reviewed by the FDA, the European Food Safety Authority, and few other regulatory agencies across the globe. Based on hundreds of scientific studies, these agencies conclude that aspartame is harmless. The exclusion is for those people who suffer from rare genetic disorder phenylketonuria (PKU); these people must wisely monitor their consumption of the vital amino acid phenylalanine, which is a constituent of aspartame. While aspartame itself does not befall in nature, it is digested to three constituents that are universal in the diet.

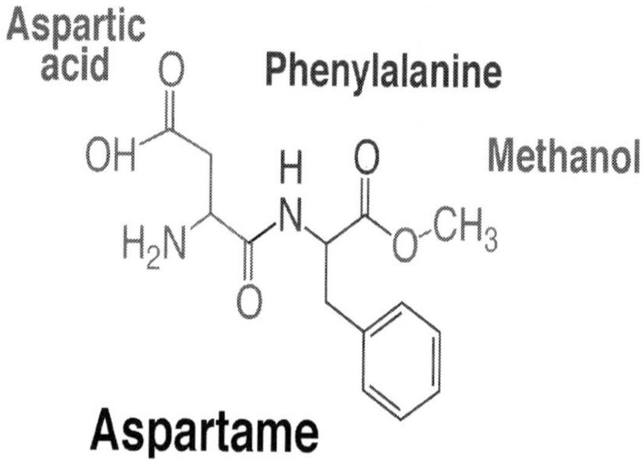

Figure 4.20: Chemical structure of aspartame

Source: http://www.sweetenerbook.com/aspartame_2.html

Aspartame is an artificial, non-saccharide or non-nutritive sweetener and is used as a sugar substitute in few foods and beverages. Aspartame is a methyl ester of the aspartic acid or phenylalanine dipeptide.

A board of experts set up by the European Food Safety Authority established in 2013 announced that aspartame is safe for human consumption at existing levels of exposure, as of 2017 sign of benefit was absent for weight loss with some data supporting weight gain and heart disease risks. Because its breakdown products of aspartame include phenylalanine, people with the genetic disorder phenylketonuria(PKU) must be attentive of this as an additional source.

Aspartame was initially sold under the brand name NutraSweet. It was originally made in 1965, and the patent deceased in 1992. It got the initial consent for use in food products by the U.S. Food and Drug Administration (FDA) in 1981.

The safety of aspartame has been the subjected to numerous political and medical controversies. Generally sold under the brand name 'Equal.' Aspartame is not recommended for canning since its sweetening properties are significantly condensed when heated. Aspartame is sold under the brand names Nutraweet® and Equal®.

- Steviol Glycosides

Steviol glycosides are the chemical compounds accountable for the sweet taste of the leaves of the South-American plant Stevia rebaudiana and the foremost ingredients of numerous sweeteners marketed under the generic name stevia and some trade names. They also happen in the associated species Stevia phlebophylla and in the plant Rubus chingii. Steviol glycosides from Stevia rebaudiana have been testified to be between 30 and 320 times sweeter than that of sucrose, though there is some discrepancy in the practical literature about these numbers.

Steviol Glycosides are heat-stable, pH-stable, and do not ferment. Moreover, they do not persuade a glycemic response when consumed, because humans cannot absorb stevia. This makes them attractive as natural sugar alternative for diabetics and on carbohydrate-controlled diets. Steviol glycosides arouse the insulin secretion through potentiation of the β-cell, stopping high blood glucose after a meal. The acceptable daily intake (ADI) for steviol glycosides, stated as steviol counterparts, has been recognized to be 4 mg/kg body weight/day, and is grounded on no experimental effects of a 100-fold higher dose in a rat study.

In 1985, scientists stated that the two of the Steviol glycosides: stevioside and rebaudioside. They are mutagen and have negative proceedings. But this proposal was criticized with contrast that even distilled water can be mutagen. In several years of studies, this product displayed both bad and good results. Bad that some searches presented steviol and stevioside to be weak mutagens. But maximum study showed absence of stevia side effects.

In 2008, 14 of 16 studies showed no genotoxic of stevioside and 11 of 15 studies showed no genotoxic of steviol and study of rebaudioside. There is no sign that this herb can source birth mutation or cancer. All searches showed that this bush does not have any negative proceedings. It only helps against metabolic, diabetes, recover insulin production, lessen hypertension and numerous other good effects.

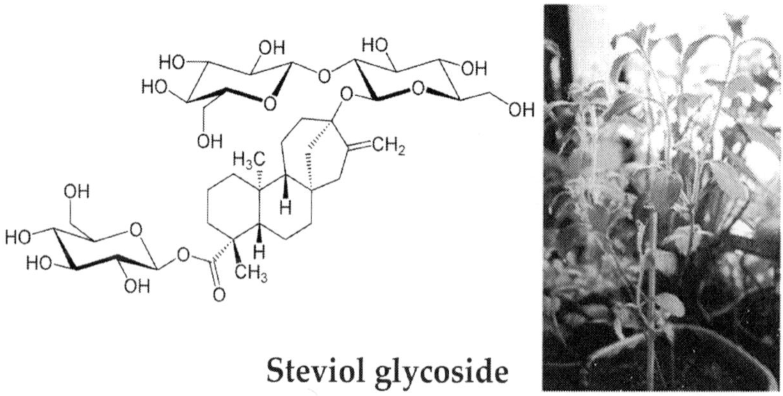

Figure 4.21: Chemical structure and plant of steviol glycosides

Source: http://chemicalsareyourfriends.com/sliders/synthetic-sweeteners-pt-2-sucralose-neotame-saccharin-and-special-natural-guest-stevia/

- Thaumatin

Thaumatin is a protein isolated from katemfe fruit, Thaumatococcus daniellii Benth, also known as "the miraculous fruit of the Sudan." Katemfe fruit is common in the West African rain forest zone. This fruit comprises of 1-3 black seeds enclosed by a gel, and covered with a membranous sac, the aril, which encompasses the sweet material. This is used as a sweetener in cooking, in flavourflavoring palm wine, and in production confections for children. Thaumatin is sweet, with a slow beginning, lasting sweetness and a licorice after-taste. Sweetness strength comparative to sucrose is about 2000x on a weight basis. Thaumatin is a protein, therefore it has four calories per gram. Since it has such high effectiveness comparative to sucrose, the quantity used is tremendously small, so it does not offer assessable caloric value at use levels in foods. Thaumatin is highly water-soluble and heat-stable. In the United States, thaumatin has GRAS position as a flavourflavor enhancer; and it does not have approval as a sweetener. JECFA and EFSA consider thaumatin harmless for use as a sweetener. Thaumatin is processed just like any other protein, thus there is no worry about toxicity. Thaumatin is stated to be allergenic, but less than house dust mites. Due to its deliberate onset and licorice-like off taste, thaumatin is questionable to be used as a sole sweetener in foods or beverages.

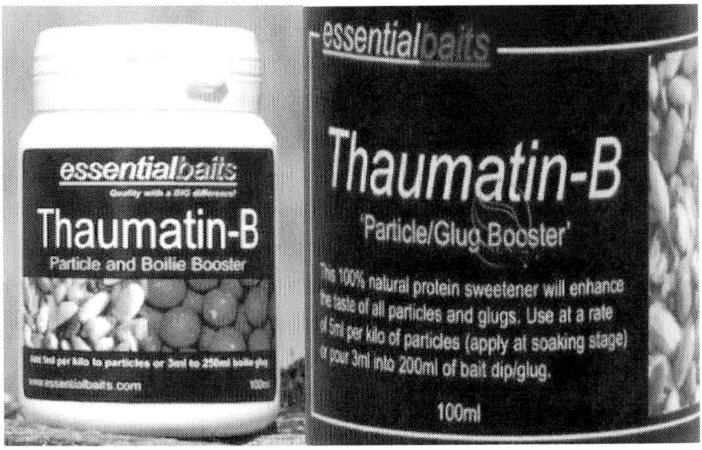

Figure 4.22: Commercially sold thaumatin with the label on the bottle

Source: http://carponline.ro/recenzii/thaumatin-b-88080

Carbohydrates

- Isomaltulose

Isomaltulose is a naturally occurring disaccharide, and is composed of alpha-1,6-linked glucose and fructose. Commercial isomaltulose is formed from sucrose by enzymatic redisposition and has been used as a sugar in Japan since 1985. It is principally appropriate as a non-cariogenic sucrose substitute and is favourablefavorable in products for diabetics and prediabetic dispositions. In-vivo studies with rats and pigs specify that isomaltulose is entirely hydrolysedhydrolyzed and absorbed in the small intestine. This is reinforced by in-vitro studies screening that intestinal disaccharidases from several species can hydrolyze isomaltulose.

The rate of hydrolysis, though, is very slow likened with sucrose and maltose. Therefore, blood glucose and insulin levels in humans after oral administration increases slowly and reach lower maxima than subsequent sucrose administration. After absorption, fructose and glucose are processed as characteristic for these monosaccharides. From intravenous studies, it is supposed that any systemic isomaltulose would be hydrolyzed as well, or defecated in urine. In numerous sub-chronic toxicity studies, the administration of large doses of isomaltulose, did not result in hostile possessions.

Isomaltulose persuaded neither embryotoxic or teratogenic effects in rat foetuses, nor maternal toxicity at levels up to 7 g/kg body weight/ day. Isomaltulose was found to be non-mutagenic in the Ames test. As the hydrolysis in the small intestine is completed, the high levels of isomaltulose are finely endured in animals and humans. In studies with healthy and diabetic focuses high doses up to 50 grams were endured without signs of intestinal uneasiness. Based on the data studied, it is determined that the use of isomaltulose as a substitute of sugar is as safe as the use of other digestible sugars containing glucose and fructose.

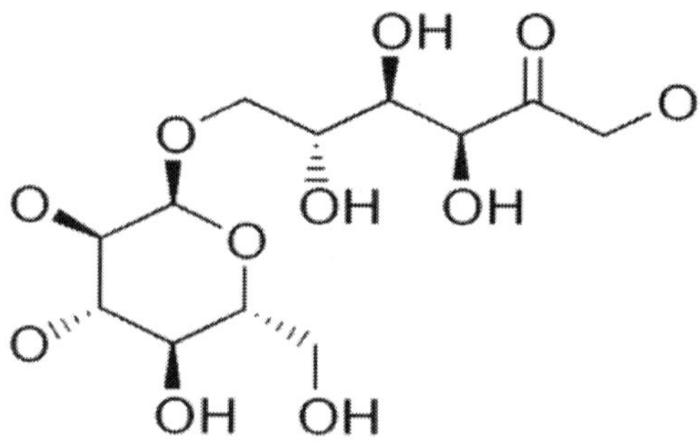

Figure 4.23: Chemical structure of isomaltalose

Source: http://cosmofarma.be/product/isomaltulose/

- Tagatose

Tagatose is a naturally occurring sugar that is obtained from lactose. Lactose or milk sugar is splited into galactose and glucose by hydrolysis. Galactose is then transformed into Tagatose by an enzymatic method. After the ultimate purification process an attractive white crystal structure with a delightful natural sugar flavourflavor is obtained. The ultimate Tagatose crystals do not contain any residual traces of lactose or glucose. Tagatose is also naturally present in many kinds of dairy products and fruit. Tagatose was formerly established by an American pharmaceutical group for the treatment of bowel cancer patients. Infact, tagatose re-establishes the intestinal flora. Later, it became seeming that it had other rewards. In 1999, The FDA (Food and Drug Administration), stated that Tagatose was an unquestionably safe ingredient to be used in foods. In 2007, this was also confirmed by the EFSA (European Food

Safety Agency). Tagatose is neither a sweetener, nor a polyol. In 2005, the Novel Food Unit classified Tagatose as a completely new food and added that it was totally safe. As it is an ingredient, it does not get an E-number.

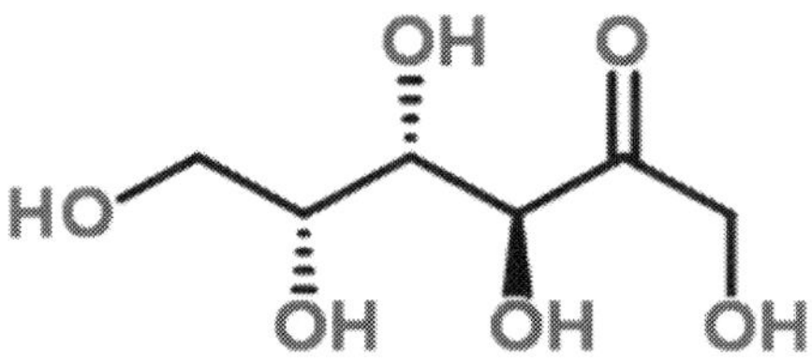

Figure 4.24: Chemical structure of tagatose

Source: http://www.chemspider.com/Chemical-Structure.83142.html

REFERENCES

1. Acsedu.co.uk. Sugar in Food Preserving -. [online] Available at: https://www.acsedu.co.uk/Info/Alternative-Living/Self-Sufficiency/Sugar-in-Food-Preserving.aspx [Accessed 29 Jul. 2017].
2. AllAboutFeed. (2014). Research: Feed flavours & sweeteners market trends. [online] Available at: http://www.allaboutfeed.net/Feed-Additives/Articles/2014/5/Research-Feed-flavors--sweeteners-market-trends-1531122W/ [Accessed 29 Jul. 2017].
3. Andrews, R. All About "Natural" Sweeteners | Precision Nutrition. [online] Precision Nutrition. Available at: http://www.precisionnutrition.com/all-about-natural-sweeteners [Accessed 29 Jul. 2017].
4. Bauman, D. (2011). A Simple Introduction to Natural Sweeteners and How to Use Them - My Humble Kitchen. [online] Myhumblekitchen.com. Available at: http://www.myhumblekitchen.com/2011/12/a-simple-introduction-to-natural-sweeteners-and-how-to-use-them/ [Accessed 29 Jul. 2017].

5. Cannold, J. (2016). Artificial vs. Natural Sweeteners. Which are Better? [online] Heather Bauer Nutrition - The Food Fix. Available at: http://www.heatherbauer.com/articles/natural-vs-artificial-sweetners [Accessed 29 Jul. 2017].
6. Home Food Preservation (Penn State Extension). (2012). Sugars and Non-Nutritive Sweeteners in Food Preservation (Home Food Preservation). [online] Available at: http://extension.psu.edu/food/preservation/news/2012/sugars-and-non-nutritive-sweeteners-in-food-preservation [Accessed 29 Jul. 2017].
7. Ingham, B. (2014). Safe & Healthy: All About Sugar Substitutes. [online] Safe and Healthy: Preserving Food at Home. Available at: http://fyi.uwex.edu/safepreserving/2014/05/20/safe-healthy-all-about-sugar-substitutes/ [Accessed 29 Jul. 2017].
8. Mains, S. Natural vs. Artificial Sweeteners - Plantbased Pixie. [online] Plantbased Pixie. Available at: http://www.plantbased-pixie.com/natural-vs-artificial-sweeteners/ [Accessed 29 Jul. 2017].
9. Mayo Clinic. (2015). Pros and cons of artificial sweeteners. [online] Available at: http://www.mayoclinic.org/healthy-lifestyle/nutrition-and-healthy-eating/in-depth/artificial-sweeteners/art-20046936 [Accessed 29 Jul. 2017].
10. Nutrients Review. (2016). Tagatose Calories, Glycemic Index, Health Benefits, Side Effects. [online] Available at: http://www.nutrientsreview.com/carbs/tagatose.html [Accessed 29 Jul. 2017].
11. Polyols-eu.com. Isomalt | Polyols EU. [online] Available at: http://www.polyols-eu.com/isomalt [Accessed 29 Jul. 2017].
12. Sugar-and-sweetener-guide.com. Mannitol. [online] Available at: http://www.sugar-and-sweetener-guide.com/mannitol.html [Accessed 29 Jul. 2017].
13. Tagatose, what is it? [Blog] Damhert nutrition. Available at: http://www.damhert.be/en/blog/artikel/13 [Accessed 29 Jul. 2017].
14. Walters, D. (2013). The Sweetener Book. [online] Sweetenerbook.com. Available at: http://www.sweetenerbook.com/isomalt.html [Accessed 29 Jul. 2017].

CHAPTER 5

AMINO ACIDS AND ORGANIC ACIDIC COMPOUNDS IN FOOD AND FEED PRODUCTION

"Food is not simply organic fuel to keep body and soul together, it is a perishable art that must be savoured at the peak of perfection."

— E.A. Bucchianeri, Brushstrokes of a Gadfly

INTRODUCTION TO AMINO ACIDS AND ORGANIC ACIDIC COMPOUNDS

The amino acid business is a multi-billion-dollar initiative. Amino acids are used as animal feed additives (lysine, methionine, threonine), in human food (leucine, isoleucine, lysine, etc.), flavourflavor enhancers (monosodium glutamic, serine, aspartic acid) and as special nutrients in the medical ground. Glutamic acid, lysine and methionine account for the mainstream, by weight, of amino acids retailed. Lysine and Glutamic acid are made by fermentation; Methionine is made by chemical synthesis. The chief producers of amino acids are grounded in Japan, the US, South Korea, China and Europe. Several microbe-based productions have their roots in traditions that go back hundreds to thousands of years. The amino acid industry has its backgrounds in food preparation practices in

Japan. Seaweeds is in use for eras there and in other Asian countries as a flavourflavoring ingredient.

Amino acids are serious to life, and have countless roles in metabolism. One principally significant role is to aid as the building blocks of proteins, which are linear chains of amino acids. Amino acids can be associated together in fluctuating sequences to form a massive variety of proteins.

Amino acids are used for a diversity of applications in manufacturing, but their foremost use is as additives to animal feed. It is necessary, since numerous of the bulk components of these feeds, such as soybeans, they either have low levels or lack some of the vital amino acids: Lysine, methionine, threonine, and tryptophan are of maximum importance in the manufacture of these feeds. The manufacturing application of amino acids for feed has a virtually 40-year history. In the late 1950s and 1960s, DL-Methionine, formed by chemical synthesis, commenced discovering its way into poultry feed. Production of L-Lysine by fermentation began in Japan in the 1960s. In addition to DL-Methionine and L-Lysine, other amino acids like- HCl, L-Threonine and L-Tryptophan were announced in the late 1980s. With advancement in biotechnology, the cost of manufacture of each amino acid has been significantly reduced, which has been one of the crucial influences in the growth of use of amino acids in animal feed.

Organic acids and their derivatives are often used in beverage, food, and feed production. Role of acidic additives is to act like buffers to control acidity, antioxidants, preservatives, flavourflavor enhancers, and sequestrants. Advantageous effects on animal health and development performance have been detected when using acidic elements as feed additives. Organic acids are being classified in groups according to their chemical structure. Each group of organic acids has its own definite property and is used for diverse applications. Organic acids, like- acetic acid, lactic acid, and citric acid, which have low molecular weight are part of the prime metabolism, and are habitually produced by fermentation. While other organic acids are produced more economically by chemical synthesis built on petrochemical raw materials on an industrial scale, for example- formic acid, propionic and benzoic acid.

Biotechnology-based production is of curiosity due to regulation, consumers appeal for natural ingredients, and growing environmental awareness. In the United States, bio-catalytically produced esters for food applications are labelled as natural, while similar conventional acid catalyst-based molecules cannot. Natural esters order or direct a price numerous times than that of non-natural esters.

Biotechnological ways need to be adjusted concerning raw materials and yield, microorganisms, and recovery approaches. New bioprocesses are being established for organic acids, which are now commercially bent by chemical synthesis. Additionally, new organic acids that can be produced with biotechnological approaches are under examination for food applications.

AMINO ACIDS IN HUMAN AND ANIMAL NUTRITION

Amino acids are the small organic molecules which plays an important role in living organisms, some of which are listed below-

- Amino acids are the chief building blocks of proteins, Nature's most miscellaneous biomolecules.
- They assist as the precursors for various biologically active molecules, like neurotransmitters (e.g. dopamine, serotonin, GABA, epinephrine), local mediators (e.g. the allergy mediator histamine), energy-related metabolites (e.g. creatine, citrulline, carnitine), oxygen-binding molecule 'heme', and DNA bases called purines.
- They also assist as an energy source in continued fasting, diabetes, and when taking the diet rich in proteins.
- Some of the amino acids act as the regulator of gene expression and cellular signaling. This effects multiple physiological processes which are related to growth, maintenance, reproduction, and immunity.

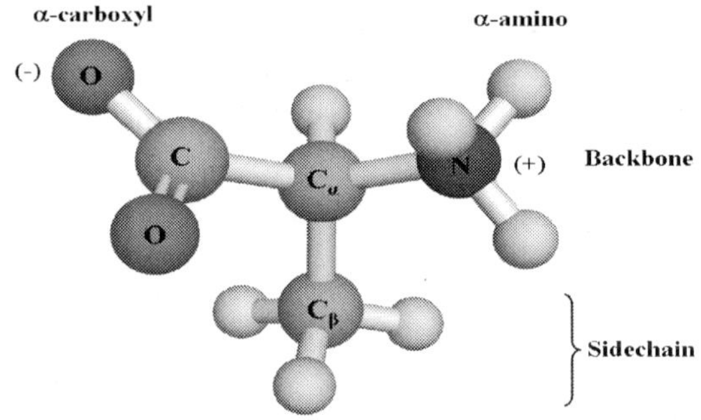

Figure 5.1: General structure of amino acid

Source: https://amit1b.wordpress.com/the-molecules-of-life/about/amino-acids/

Out of the 20 amino acids that are present, nine are essential amino acids and 11 of them are non-essential amino acids. Below are a list of the nine essential amino acids and plant-based foods that are decent sources of each. Some sources of amino acids, such as chia and hemp seeds, offers all essential amino acids, making them a whole protein. The list of essential amino acids includes: Histidine, isoleucine, leucine, methionine, cystine, phenylalanine, tyrosine, threonine, tryptophan, and valine.

Essential	Conditionally Non-Essential	Non-Essential
Histidine	Arginine	Alanine
Isoleucine	Asparagine	Asparatate
Leucine	Glutamine	Cysteine
Methionine	Glycine	Glutamate
Phenylalanine	Proline	
Threonine	Serine	
Tryptophan	Tyrosine	
Valine		
Lysine		

Figure 5.2: The 20 essentials, conditionally non-essential, and non-essential amino acids

Source: https://www.fastbleep.com/biology-notes/40/753

Leucine

Leucine is one of the finest essential amino acids for motivating muscle strength and growth, and denoted as a BCAA (brand-chain amino acid). Leucine helps in controlling the blood sugar by regulating insulin into the body during and after exercise and can even help avert and treat depression by the method it acts on neurotransmitters in the brain. Decent plant-based sources include: seaweed, pumpkin, peas and pea protein, whole grain rice, sesame seeds, watercress, turnip greens, soy, sunflower seeds, kidney beans, figs, avocados, raisins, dates, apples, blueberries, olives and even bananas.

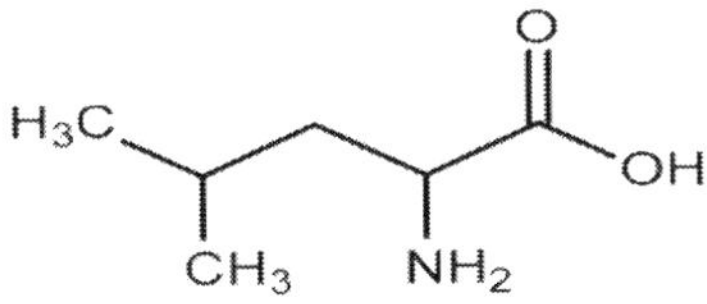

Figure 5.3: Structure of leucine

Source: http://www.i-supplements.com/leucine.html

Isoleucine

Isoleucine is one more BCAA alike leucine, though with a few dissimilar responsibilities. It is an isolated arrangement of leucine that precisely assists the body to produce energy and haemoglobin. It is also dynamic assisting in nitrogen growth within the muscle cells, particularly in children. Plant-based sources include: rye, soy, cashews, almonds, oats, lentils, beans, brown rice, cabbage, hemp seeds, chia seeds, spinach, pumpkin, pumpkin seeds, sunflower seeds, sesame seeds, cranberries, quinoa, blueberries, apples, and kiwis.

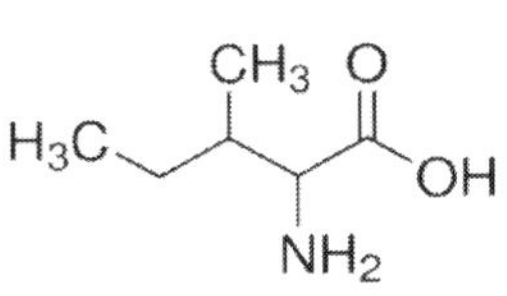

Figure 5.4: Structure of isoleucine

Source: http://www.sigmaaldrich.com/catalog/product/sigma/i6268?lang =en&region=IN

Lysine

Lysine is accountable for appropriate growth and in the production of carnitine. Lysine also helps the body to absorb calcium for further bone strength and helps in collagen production. It is vital to take sufficient of this amino acid since shortage can lead to nausea, depression, fatigue, muscle depletion and osteoporosis.

Decent plant-based sources of lysine comprise: beans (the best), watercress, hemp seeds, chia seeds, spirulina, parsley, avocados, soy protein, almonds, cashews, and some legumes with lentils and chickpeas being two of the best.

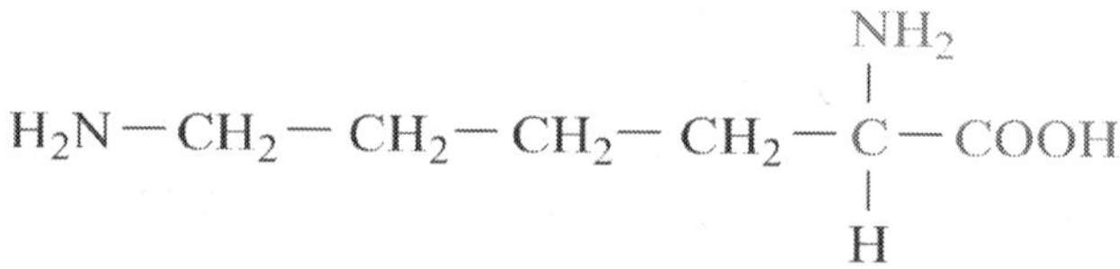

Figure 5.5: Structure of lysine

Source: http://glossary.periodni.com/glossary.php?en=lysine

Methionine

Methionine supports in the formation of cartilage in the bod sulphur SulphurSulfur is a mineral, crucial to the production of bone cartilage and no other amino acids sulphursulfur apart from methionine. People who don't eat pl sulphursulfur-containing foods to produce methionine in the body they may suffer from arthritis, damaged tissue, and poor healing. Methionine also helps in the production of muscle development and formation of creatine, desirable for ideal cellular energy.

Good plant-based sou sulphursulfur are: sunflower seed butter and sunflower seeds, hemp seeds, chia seeds, Brazil nuts, oats, seaweed, wheat, figs, whole grain rice, beans, legumes, onions, cacao, and raisins.

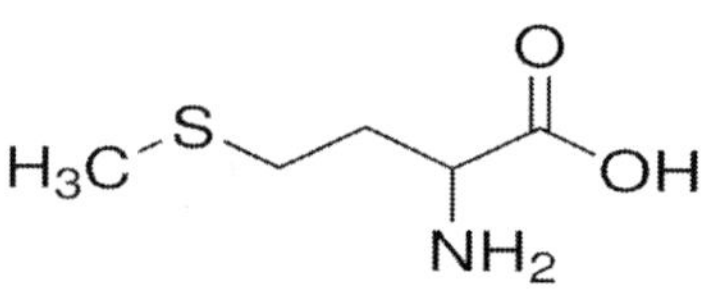

Figure 5.6: Structure of methionine

Source: http://www.mpbio.com/product.php?pid=02190955&country=99

Phenylalanine

Phenylalanine comes in three different forms: L-phenalynaline, which is a natural form found in protein; D-phenalynaline, which is a form produced by a laboratory, and DL-phenalynaline, which is a mixture of both forms. Always eat food-based sources before picking supplemented or improved food products with a lab-derived version of this amino acid.

Phenylalanine is significant in the body because it turns into tyrosine once it is ingested, which is another amino acid that is desired to make proteins, brain chemicals, and thyroid hormones. Not gaining plenty of this amino acid can result in brain fog, deficiency of energy, depression, lack of appetite, or memory problems. Good sources include: spirulina and other seaweed, pumpkin, beans, rice, avocado, almonds, peanuts, quinoa, figs, raisins, leafy greens, most berries, olives, and seeds.

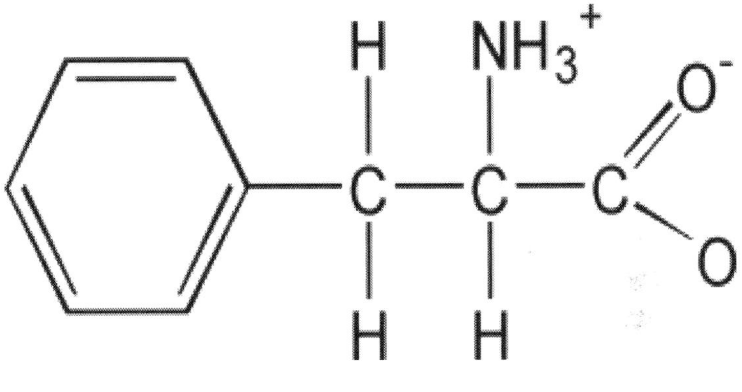

Figure 5.7: Structure of phenylalanine

Source: https://smartdrugsforcollege.com/phenylalanine-side-effects/

Threonine

Threonine helps in making a healthy immune system, heart, liver, and central nervous system health. Threonine also helps in maintaining an equilibrium of proteins within the body to contribute to overall repair, energy, and growth. This amino acid also supports the body's connective tissues and joints in decent health by producing glycine and serine in the body, the two most vital amino acids required for healthy bones, skin, hair, and nails. In the liver, it benefits with fatty acid digestion to avert fatty acid build-up and liver failure. The maximum sources of this amino acid are: watercress and spirulina, pumpkin, leafy greens, hemp seeds, chia seeds, soybeans, sesame seeds, sunflower seeds and sunflower butter, almonds, avocados, figs, raisins, quinoa, and wheat. Sprouted grains are also outstanding sources of this amino acid.

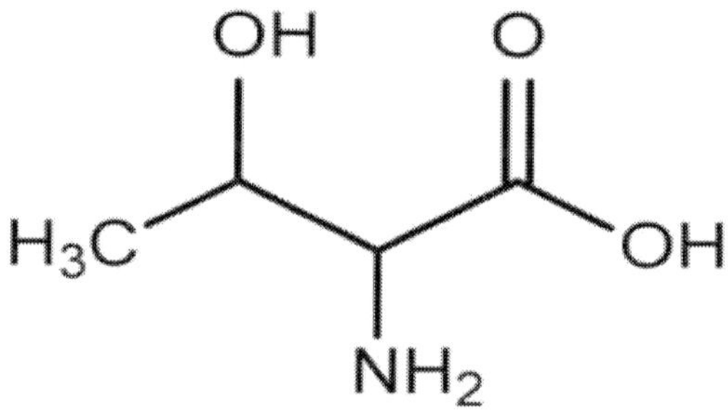

Figure 5.8: Structure of threonine

Source: http://www.naturestrailwellness.com/health/the-amino-acid-threonine/

Tryptophan

Tryptophan is recognized as the relaxing amino acid, tryptophan is vital to a healthy nervous system and brain health, along with sleep, muscle development and repair, and complete neurotransmitter function. It is one of the most protruding amino acids found in turkey, milk, and cheese that cause these foods to make the consumer feel sleepy and relaxed. Tryptophan also changes to serotonin when in the brain, which generates a happy feeling tied to lower levels of stress and unhappiness. Animal foods endorse swelling and so tons of plant-based sources are there instead.

Some of the plant-based sources that include high amounts of tryptophan are: oats and oat bran, seaweed, hemp seeds, chia seeds, spinach, watercress, soybeans, pumpkin, sweet potatoes, parsley, beans, beats, asparagus, mushrooms, all lettuces, leafy greens, beans, avocado, figs, winter squash, celery, peppers, carrots, chickpeas, onions, apples, oranges, bananas, quinoa, lentils, and peas.

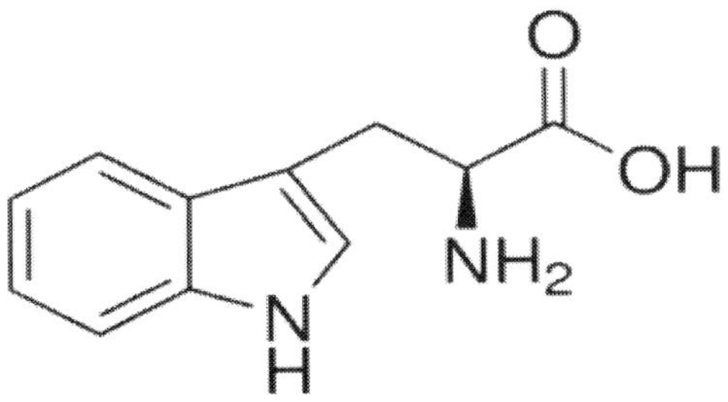

Figure 5.9: Structure of tryptophan

Source: *http://www.sigmaaldrich.com/catalog/product/sial/t2610000?lang=en &region=IN*

Valine

Valine is another BCAA required for optimal muscle development and repair. It is also accountable for stamina and the complete maintenance of good muscle health.

High plant-based sources of valine include: beans, spinach, legumes, broccoli, sesame seeds, hemp seeds, chia seeds, soy, peanuts, whole grains, figs, avocado, apples, sprouted grains and seeds, blueberries, cranberries, oranges, and apricots.

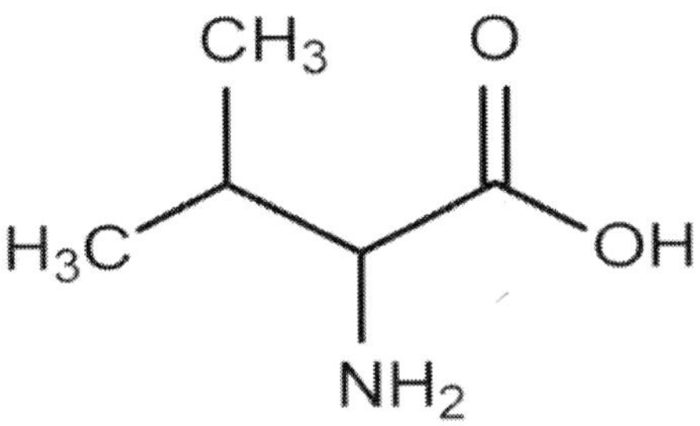

Figure 5.10: Structure of valine

Source: http://www.planetayurveda.com/library/valine

Histidine

Histidine, amino acid supports in transporting neurotransmitters to the brain and benefits overall muscle health within each muscle cells. It also helps in detoxifying the body by producing red and white blood cells desired for complete health and immunity. Not gaining plenty of histidine may result in arthritis, sexual dysfunction, and even deafness. It can also make the body more vulnerable to the AIDS virus. Good plant-based sources of histidine are: rice, wheat, rye, seaweed, beans, legumes, cantaloupe, hemp seeds, chia seeds, buckwheat, potatoes, cauliflower, and corn.

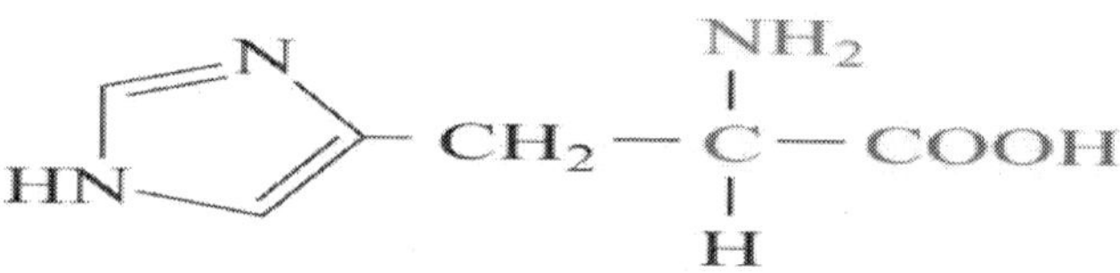

Figure 5.11: Structure of histidine

Source: http://glossary.periodni.com/glossary.php?en=histidine

Amino Acid	Main Food Sources
Histidine	soy protein, eggs, parmesan, sesame, peanuts
Isoleucine	eggs, soy protein & tofu, whitefish, pork, parmesan
Leucine	eggs, soy protein, whitefish, parmesan, sesame
Lysine	eggs, soy protein, whitefish, parmesan, smelts
Methionine	eggs, whitefish, sesame, smelts, soy protein
Cysteine	eggs, soy protein, sesame, mustard seeds, peanuts
Phenylalanine	eggs, soy protein, peanuts, sesame, whitefish
Tyrosine	soy protein, eggs, parmesan, peanuts, sesame
Threonine	eggs, soy protein, whitefish, smelts, sesame
Tryptophan	soy protein, sesame, eggs, winged beans, chia seeds
Valine	eggs, soy protein, parmesan, sesame, beef

Figure 5.12: Essential amino acids and its main sources

Source: https://amit1b.wordpress.com/the-molecules-of-life/about/amino-acids/

Essential amino acids in food and feed

Amino acids for feed plays a very vital role in improving the efficacy of protein utilization in animal feeding. Some of the amino acids used in animal feed production are-

L-Threonine

The chemical formula of L-threonine is $C_4H_9NO_3$.

The molar mass of L-threonine is 119.1192 g/mol.

L-threonine is classified as an essential amino acid. Humans do not manufacture L-threonine, so it must be gained through the diet. As there is no manufacturing of this amino acid in the human body it comes under the category of essential amino acids and needs to be taken in certain specific amount through diet. The male adult's daily necessity is 7 mg per kg of body weight. Maximum people will get adequate L-threonine through the diet, but as threonine content in grains is low, vegetarians are more likely to have deficiencies and should think of some supplementation.

L-threonine is a crucial constituent of collagen, elastin and enamel proteins. It helps in appropriate fat metabolism in the liver; nutrients are more voluntarily absorbed when threonine is present. L-threonine also supports the digestive and intestinal tract to function more effortlessly. Specialists trust that L-threonine improves the immune system by helping in the production of antibodies. L-threonine is mass-produced by fermentation from carbohydrate sources. L-threonine is mass-produced under strict cGMP conditions. No resources of animal origin are used in the making of L-threonine.

L-threonine meets USP, EP and JP specifications and has a Drug Master File on top of the FDA. L-threonine is mass-produced under strict cGMP conditions. No resources of animal origin are used in the making of L-threonine. L-threonine meets USP, EP and JP specifications and has a Drug Master File on top of the FDA.

L-Threonine is an important amino acid that is added to the diet of pigs and poultry to accurately match the balanced amino acid necessity of the animals. In repetition, L-Threonine is the second restrictive amino acid for pigs and the third for poultry. Even in the low protein diets with balanced L-Threonine content, L-Threonine can improve growth performance and reduce the excretion of nitrogen. L-threonine is cost-effective and environment friendly.

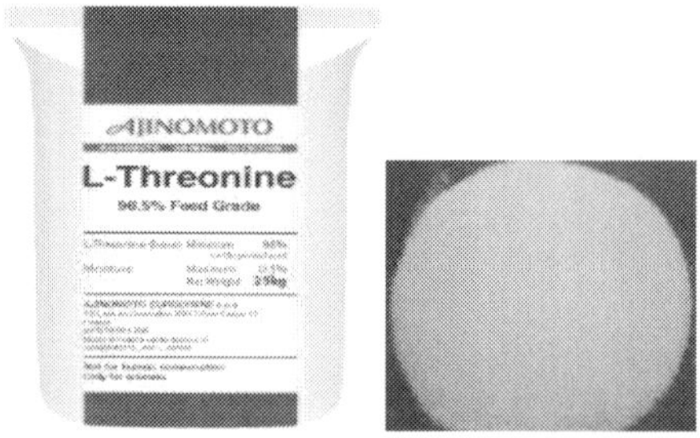

Figure 5.13: L-threonine

Source: http://www.eksmatrade.lt/amino-acids/

L-Lysine

L-lysine is an essential amino acid. Amino acid shortages cause a degradation of inner cells and often leads to severe problems, therefore it is vital to take sufficient amount in the diet. Essential amino acids are most habitually stereoisomers, in a sense that they exist in two variations that are alike to one another as mirror images. The two forms are D- and L- forms of these amino acids, and the L- form is used in protein synthesis and thus the form found in food and supplements. There are various unbelievable L-lysine benefits, starting from its use as a common treatment for cold sores to a possible anti-anxiety supplement.

L-lysine can be taken orally also but is best absorbed in the body when consumed through food. It is found in huge quantities in several types of meat, beans, cheeses, and eggs. L-lysine is very imperative in the formation of carnitine, which converts fatty acids into energy and drops cholesterol levels. L-lysine also appears to play an important part in absorbing calcium and supports the body to form collagen, which helps in the growth and maintenance of bones and connective tissue. As a natural disease-fighting agent, L-lysine welfares the human body in a diversity of conducts, some of which are only freshly undergoing study. Benefits of L-lysine are-

- May Decrease Outbreak and Frequency of Herpes Viruses.
- May help in the treatment of cancer.

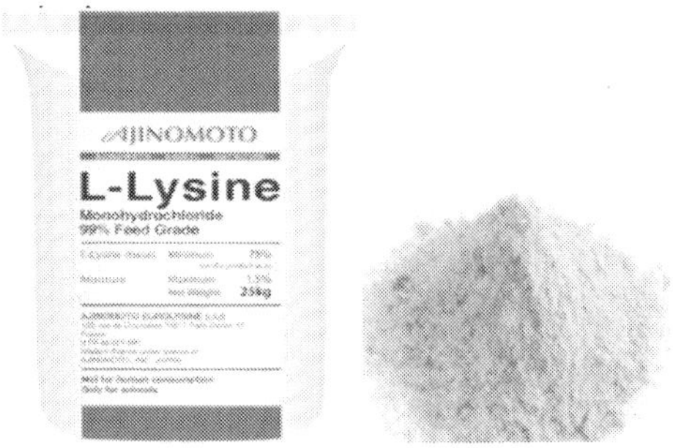

Figure 5.14: L-lysine

Source: http://www.eksmatrade.lt/amino-acids/

Taurine

Taurine is a kind of amino acid found throughout the body. It is mainly concentrated in the brain, eyes, heart and muscles. Unlike many other amino acids, it is not used to build proteins in the body. It is classified as "conditionally essential" amino acid. Our body can produce approximate quantity of taurine, and it is also found in some foods. However, certain persons may get benefitted by taking a supplement. Individuals with specific illnesses or diseases, like heart issues or diabetes, may also profit from additional taurine intake. The chief source of taurine is animal foods like- meat, fish and dairy. While some vegetarian foods deliver it in lesser quantities, it is unlikely that these will deliver adequate amounts to enhance the body's levels. Taurine is also added to soda and energy drinks, that may offer upto 600-1,000 mg in an 8-ounce serving. However, it is not suggested to drink soda or energy drinks due to other constituents that are injurious. The form of taurine that is used in supplements and energy drinks is usually made unnaturally and not the derivative of animal sources. It is therefore appropriate for vegetarians. An average diet will deliver about 40–400 mg of taurine per day, but studies have used supplements with up to 400–6,000 mg per day.

Taurine is found in numerous organs, and its benefits are extensive. Its direct roles are in:

- Sustaining appropriate hydration and electrolyte stability in the cells.

- Creating bile salts, which play a significant part in digestion.
- Regulating minerals, like- calcium within cells.
- Supporting the overall function of the central nervous system and eyes.
- Regulating immune system health and antioxidant function.

Taurine (chemical name 2-aminoethansulphoric acid) is a naturally happening beta-amino acid in mammals. It is the building block of a significant constituent of bile acid and extremely concentrated in central nervous system, skeletal muscle, heart muscle and eyes, specifically in the brain and heart muscle. In mammals, it acts as conjugator of bile acid, aids eye health and vision, and plays detoxification, osmoregulation, and membrane stabilization, and hence, ensures the regular reproduction, growth and maintenance. As Taurine could be produced in the body from other components, it is commonly documented as non-essential amino acid for mammals. It is extensively functional in infant food, energy fortified beverage and pet foods, where the synthesis of Taurine is inadequate and the dietary supplementation is obligatory.

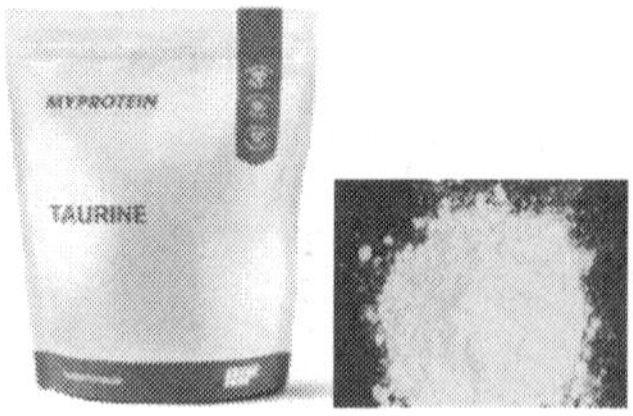

Figure 5.15: Taurine

Source: http://www.eksmatrade.lt/amino-acids/

L-Tryptophan

L-Tryptophan is also an essential amino acid that needs to be taken through external diet as body cannot produce it. It is used for insomnia, depression, anxiety, facial pain, premenstrual dysphoric disorder (PMDD)- a severe form of premenstrual pain, smoking cessation, bruxism, attention deficit-hyperactivity disorder (ADHD), and also to boost athlete's performance. It is naturally found in plant and animal protein. L-tryptophan is essential for the normal development and functioning of various organs of the body. Once L-tryptophan is absorbed, body converts it into 5-HTP (5-hydroxytryptophan), and then to serotonin. Serotonin helps in transmitting signals between the nerve cells. It also leads to narrowing of the blood vessels and any changes in serotonin's level in the brain can alter mood.

L-Tryptophan 98% Feed Grade" is used as a feed supplement to meet the requirements of digestible tryptophan in animals. Predominantly it is appropriate for post-weaning piglets, since it supports to enhance feed intake and confines the undesirable effect of unfavorable health environment, and ultimately leading to enhanced growth rate anductivity.

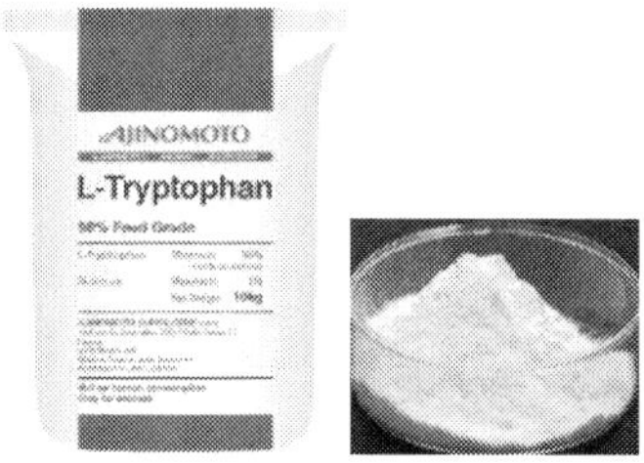

Figure 5.16: L-tryptophan

Source: http://www.foodchemadditives.com/products/l-tryptophan

MaximisingMaximizing feed intake is the biggest encounter in post-weaning piglets. At both physiological stages, enhancing a lacking diet with tryptophan has been verified to increase feed intake. This effect hence shows that this amino acid is involved in appetite regulation in animals. In the post-weaning phase, diets play a role in growth promotion, feed intake, keep of a decent sanitary status and evasion of stress. The several tryptophan purposes are linked to each of these features, thus it is significant to use a satisfactory tryptophan level in piglet diets.

PRODUCTION AND PROPERTIES OF ORGANIC ACIDS

An organic acid is nothing but an organic compound with acidic properties intact. The most occurring organic acids are the carboxylic acids, whose acidity is related with their carboxyl group –COOH. Sulphonic acids, comprising the group $-SO_2OH$, are comparatively stronger acids.

Alcohols, with –OH, can also act as acids but they are typically very weak. The relative stability of the conjugate base determines its acidity. Other groups can also confer acidity, but typically very weakly, like- the thiol group –SH, the enol group, and the phenol group. In biological structures, organic compounds containing these groups are commonly denoted as organic acids. Simple organic acids such as formic acid or acetic acid are used for oil and gas well stimulation actions. These organic acids are very less reactive with metals than those of strong mineral acids, example- hydrochloric acid (HCl) or mixtures of HCl and hydrofluoric acid (HF). Due to this reason, organic acids are used at high temperatures or when extended interaction periods between acid and pipe are required. The conjugate bases of organic acids like citrate and lactate are frequently used in biologically-compatible buffer solutions. Citric acid and oxalic acids are often used as rust removals. They can dissolve the iron oxides, as acids, but without damaging the base metal as stronger mineral acids do. In the dissociated form, they may chelate the metal ions, serving to speed removal. Biological systems produce numerous complex organic acids like L-lactic, citric, and D-glucuronic acids that contain hydroxyl groups or carboxyl groups. Human blood and urine cover these plus organic acid degradation products of amino acids, neurotransmitters, and intestinal bacterial action on food apparatuses. For example- alpha-ketoisocaproic, vanilmandelic, and D-lactic acids, derived from catabolism of L-leucine and epinephrine by human tissues and catabolism of dietary carbohydrate by intestinal bacteria.

Organic acids are play a vital role in food preservation because of their effects on bacteria. The basic principle on the mode of action of organic acids is that non-dissociated organic acids can penetrate the bacteria cell wall and disrupt the normal physiology of specific types of bacteria that are pH-sensitive, which means that they cannot tolerate an inclusive internal and external pH gradient. Several examples of those bacteria are Escherichia coli, Salmonella spp., C. perfringens, Listeria monocytogenes, and Campylobacter species. Upon passive diffusion of organic acids, where the pH is somewhere near or above neutrality, the acids will dissociate and subordinate the bacteria internal pH, leading to conditions that will damage or stop the growth of bacteria. Afterwards, the anionic portion of the organic acids, which cannot escape the bacteria in its dissociated form, will accumulate within the bacteria and interrupt various metabolic functions, leading to increase in the osmotic pressure, incompatible with the persistence of the bacteria. It has been well confirmed that the state of the organic acids is enormously significant in determining their capacity to prevent the growth of bacteria. Lactic acid and its salts sodium lactate and potassium lactate are extensively used as antimicrobials in food products, particularly in meat and poultry like ham and sausages.

Organic acids have been used effectively in pig production for about 25 years and more. Though less research has been done in poultry, organic acids have been found to be active in poultry production. Organic acids are commonly dispersed in nature as normal elements of plants or animal tissues. They are also designed through microbial fermentation of carbohydrates mostly in the large intestine. They are sometimes found in their sodium, potassium, or calcium salts, or even in stronger double salts. Organic acids that are added to feeds must be protected to avoid their dissociation in the crop and in the intestine and spread far into the gastrointestinal tract, where the bulk of the bacteria population is positioned. Using organic acids in poultry and pigs, one can imagine an enhancement in performance alike the growth of antibiotic growth promoters, without the public health concern, a preventive effect on the intestinal difficulties such as necrotic enteritis in chickens and Escherichia coli infection in young pigs.

Organic acids and its uses

The organic acids and their derivatives spreads to an extensive variety of substances. They are used in all kind of chemically manufactured elements. These molecules are generally considered as irritant, the level of such irritant is decided partly by separation of acids along with the water solubility capacity. Some of these could affect in severe burning or virtually alike effects as the mineral acids.

Numerous of these carboxylic acids are of huge importance because of their use in food and beverage industries along with drugs and other industrial procedures. Some of the most common organic acids which are used in manufacturing are adipic acid, azelaic acid, Fumaric acid, maleic acid, malic acid, malonic acid, oxalic acid, succinic acid, tartaric acid and sebacic acid. The saturated long chain of the monocarboxylic acids are essentially fatty acids. These are the foremost chain resourced from several natural sources. Synthetic fatty acids could also be organized by oxidation of paraffin using metals as catalysts. These could also be arranged by alcohol oxidation in existence of alkaline potassium di chromates. The marketable naphthenic acid is derivative of cyclo-paraffins in petroleum, by oxidation. These organic acids are used in plastic making, tanning, textiles, paper, metal, pharmaceuticals, food industries, beverage making and in manufacturing of cosmetics. These organic acids are also used in manufacturing of perfuerbal insecticides, colourcolors and dyes, many lubricants and common household cleaners.

Formic acid and acetic acid

They are measured as the most important industrial chemicals amongst saturated monocarboxylic acids. Although, formic acid is primarily used in textile and leather industry, acetic acid serves as an intermediate chemical acting as a restricting agent during leather tanning, aid as solvent, and as oil well acidizer. Formic acid acts as a neutralizing agent in leather industry and as a dye exhausting agent in chrome dying. Apart from these acetic acid acts as additive in number of food products and as a dying agent in dye stuff, pharmaceutical industries and textile industries.

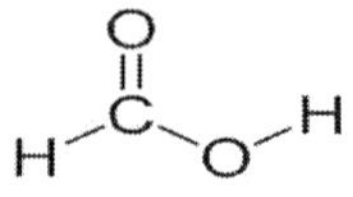

Figure 5.17: Formic acid

Source: http://registrardaily.com/2017/04/18/global-formic-acid-99-100-market-2017/

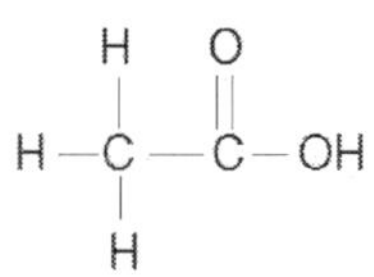

Figure 5.18: Acetic acid

Source: https://socratic.org/questions/what-is-the-oxidation-state-for-carbon-in-ch3cooh

Chloroacetic acid

The Chloroacetic acid is mainly used in pharma industries, in dye stuff and in chemical industries as a chemical intermediate. The salicylates are used as a chemical intermediate in the making of aspirin formulations, rubber industries and dye stuff industries. Benzoic acid, the ascorbates and oleates are some other valuable derivatives that are used in food, beverage and pharma industries.

Acids like palmitic acid and stearates are used in many industries, such as in soap industries, cosmetic preparation, the cleaning agents such as detergents, machinery agents such as lubricants, and hard protective coatings and other intermediate products.

- Propionic acids are used in organic synthesis, used as several fungicides and as food preserving chemical.
- Meth acrylic acid, Acrylic acid, Crotonic acid are used in the making of resins, paper, plastics and pain industries.
- Crotonic acid is also used as a softening agent in synthetic rubber manufacturing in rubber industry.
- Lactic acid is mainly used in adhesive, plastics and textiles.
- Glycolic acids are chiefly used in leather, textile, and electroplating, adhesive and metal cleaning industries.

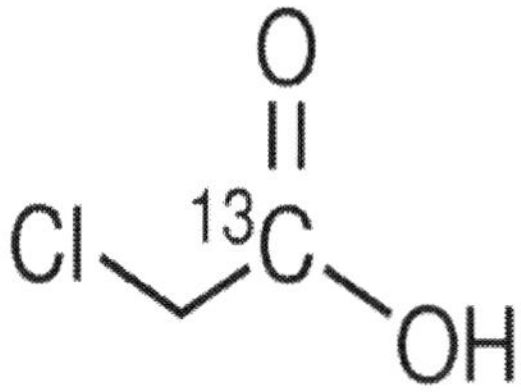

(a)

(b)

Figure 5.19: (a) Structure of chloroacetic acid, (b) chloroacetic acid powder

Source: http://www.sigmaaldrich.com/catalog/product/aldrich/488496?lang=en&region=IN

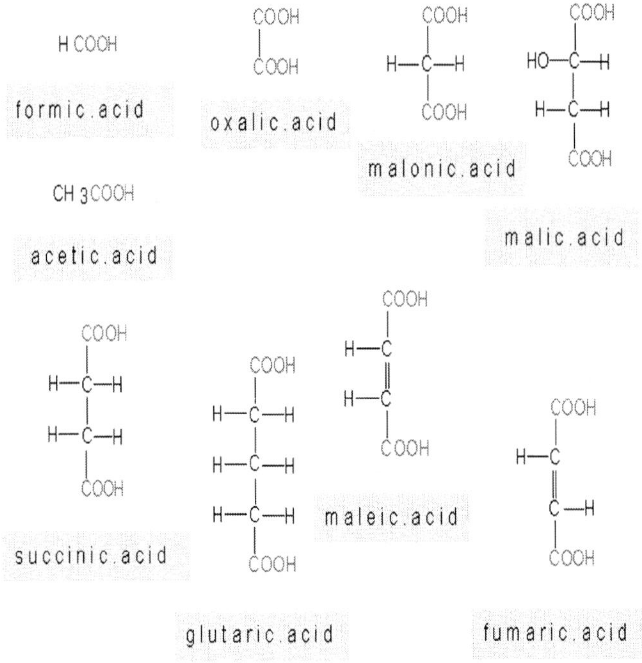

Figure 5.20: Structure or organic acids

Source: http://www1.biologie.uni-hamburg.de/b-online/e16/16e.html

Types of organic acid

Organic acids mainly are of two types. Out of the two one is derived from the -COOH group while the other is derived from the phenolic group or C6H5OHC6H5OH. The dicarboxylic acids, for example- the succinates, the fumarates and maleic acid along with tricarboxylic acid such as citrates are very valuable in food, beverage and pharma industries.

- Succinic acid is predominantly used in lacquers and dye manufacturing.
- Maleic acid is used in synthetic or artificial resin manufacturing and as a preservative for oils and fats.
- Fumaric acid is used in the manufacturing of polyester resins, several ink and organic syntheses.
- Adipic acid is used to produce nylon, and synthetic lubricants, long chains of urethanes and food acidulates.
- Oxalic acid is commonly used for cleaning, textile finish, and household metal cleaning formulations. It is also used as a bleaching agent of leather, and in syntheses.

The acids which are produced by plants and animals are named as organic acids. Various of these are found in the living things. The list below is of organic acids that are found in living things-

Table 5.1: Organic acids and their origin

Acids	**Origin**
Citric acid	Citrus fruits like oranges and lemons
Tartaric acid	Fruits like grapes and tamarinds
Ascorbic acids	Present in citrus fruits as Vitamin C
Methanoic acids	The sting of nettle, and ant sting
Oxalic acids	The juice of tomatoes
Uric acids	The urine discharge
Lactic acids	When milk goes sour

Organic acids are recurrently used in beverage, food, and feed production. Acidic additives act as buffers to regulate acidity, antioxidants, preservatives, flavor enhancers, and sequestrants. Advantageous effects on animal health and growth performance have been detected when using acidic substances as feed additives. Organic acids could be classified in clusters conferring to their chemical structure. Each group of organic acids has its particular detailed property and are used in diverse applications. Organic acids with low molecular weight, for example- acetic acid, lactic acid, and citric acid, which are part of the primary metabolism, are habitually produced by fermentation. Others are produced more economically by chemical synthesis founded on petrochemical raw materials on an industrial scale, for example- formic acid, propionic and benzoic acid.

L-ASCORBIC (VITAMIN C) AND D-ISOASCORBIC ACID

L-ascorbic acid (vitamin C) was originally isolated in 1928 and afterwards it was recognized as the long-sought antiscorbutic factor. Mechanically produced L-ascorbic acid is generally used in the feed, food, and pharmaceutical sector as nutritional supplement as well as preservative, making use of its antioxidative properties. Till today, the Reichstein–Grüssner process, which was designed in 1933, was the foremost industrial route. In this process, D-sorbitol is converted to L-ascorbic acid via 2-keto-l-gulonic acid (2KGA) which acts as crucial intermediate, using a bio-oxidation with *Gluconobacter oxydans* and some chemical steps.

Today, industrial production developments use other bio-oxidation steps with *Ketogulonicigenium vulgare* as biocatalyst to change d-sorbitol to the intermediate 2KGA without chemical steps. The enzymes involved are categorized by an extensive substrate range, but remarkable regions specificity. This confusing specificity design can be understood from the preferences of these enzymes for certain of the various isomeric structures which are the carbohydrate substrates accepted in aqueous solution.

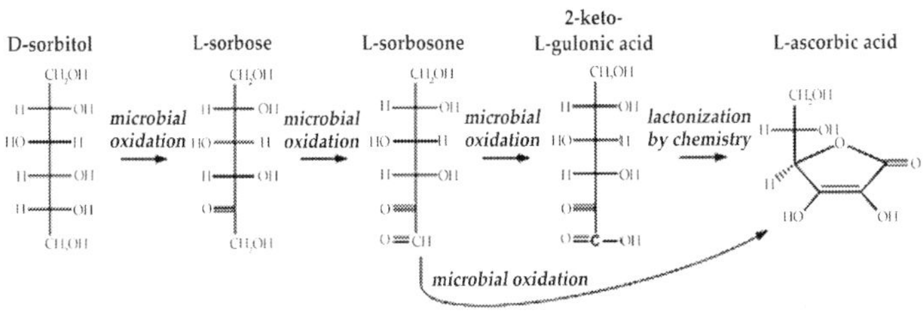

Figure 5.21: Graphical abstract of L-ascorbic acid

Source: https://link.springer.com/chapter/10.1007%2F10_2013_243

Lately, novel enzymes were recognized that produce L-ascorbic acid straight via oxidation of L-sorbosone, which is an intermediate of the bio-oxidation of D-sorbitol to 2KGA. This opens the likelihood for the shortest route from D-sorbitol to L-ascorbic acid, avoiding the requirement for chemical rearrangement of 2KGA. Analogous concepts for industrial processes apply for the production of D-isoascorbic acid, which is a C5 epimer of L-ascorbic acid. D-isoascorbic acid has the similar conformation at C5 as D-glucose and can be derived directly than L-ascorbic acid from the common carbohydrate feed stock.

L-ascorbic acid or vitamin C befalls in nutritionally momentous amounts as an accepted component of many fruits, vegetables, berries, and melons. Since it is a vitamin, it is required in the diet of all age groups. L-ascorbic acid and their sodium salt are antioxidants which are widely used as preservativlourcolor stabilizers and other related functions in many kinds of foods and beverages. Calcium ascorbate and ascorbyl palmitate, which are the derivatives of ascorbic acid with better fat solubility, also are anti-oxidants, but they do not have any substantial use in processed foods. Apart from its use in foods as antioxidants, L-ascorbic acid and its salts are also added to different foods as a source of vitamin C. These sources produce a vital amount of the total ascorbate consumption of the general population.

D-isoascorbic acid or erythorbic acid, is a stereoisomer of L-ascorbic acid, and its sodium salt, they are also the effective antioxidants and for this purpose they are used in many food products. The amounts used in 1970s were significantly less than that for the ascorbates. The vitamin action of erythorbates is only one -twentieth to that of ascorbic acid, and the anti-oxidant efficiency is no greater than that for the ascorbates. Due to this reason, it is desirable to use L-ascorbic acid rather than erythorbic acid as an antioxidant wherever possible.

The research on guinea pigs and man has committed that though erythorbic acid shares analogous absorption and tissue uptake system as ascorbic acid, still it has a little antiscorbutic activity. While struggle between ascorbic acid and erythorbic acid has been established at a biochemical level, there is no fixed evidence that such competition will produce a scorbutic state. Both short-term and long-term toxicity studies have confirmed tolerance without adverse effects for huge quantities of orally administered L-ascorbic acid, sodium L-ascorbate, and erythorbic acid in numerous species including mice, rats, guinea pigs, rabbits, and dogs. A significant number of short-term trials with human subjects ingesting 1 to 4 g of ascorbate daily have normally not exposed any damaging effects. Some human subjects have received higher amounts, upto at least 8 to 10 g per day. In most illustrations, no unfortunate fallouts have been noted. But there is noticeable scantiness of such studies that were well controlled and in which inquisitive consideration was given to likely damaging effects. In due course, such studies would be needed.

In the numerous studies on the effect of consumption of extreme quantities of ascorbates, attention has been focused on queries including oxalate excretion and renal tract stones, effects on the use of copper, iron, and other metals, requirement for vitamin B12, blood coagulation, and reproductive performance. The results specify that the tolerance to extreme quantities of ascorbic acid and its sodium salt is high. Some investigators have stated the development of dependence in animals and human after absorption of large amounts of ascorbates for prolonged time periods; still, the levels of ascorbates added to foods by 1 to 3 orders of magnitude. It is noteworthy that no data have been created regarding the likely effects of ascorbyl palmitate and calcium ascorbate in humans, and there is virtually no evidence concerning the latter in animals. Evidence concerning ascorbyl palmitate in animals is practically limited. The rare meaningful trials recommend that ascorbyl palmitate is tolerated near the similar level as ascorbic acid and sodium ascorbate. It is rational to adopt that the tolerance to calcium ascorbate is about the same as for sodium ascorbate and this is a high level.

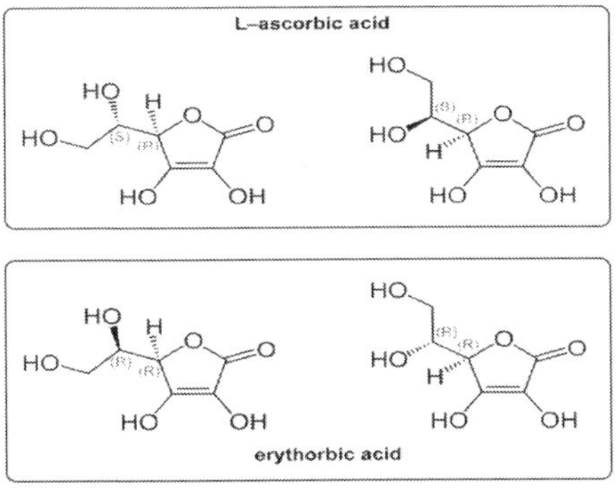

Figure 5.22: Structures of L-ascorbic acid (vitamin C) and D-isoascorbic acid (erythorbic acid)

Source: https://commons.wikimedia.org/wiki/File:Ascorbic-acid-vs-erythorbic-acid-2D-skeletal.png

REFERENCES

1. Ajiaminoscience.com. (2017). L-Threonine. [online] Available at: http://www.ajiaminoscience.com/products/manufactured_products/l-amino_acids/L-Threonine.aspx [Accessed 29 Jul. 2017].
2. Chemistry.tutorvista.com. (2017). Organic Acids - Definition, List & Uses | Chemistry@TutorVista.com. [Online] Available at: http://chemistry.tutorvista.com/inorganic-chemistry/organic-acids.html [Accessed 29 Jul. 2017].
3. Dr. Axe. (2017). Top 10 Highest L-Lysine-Rich Foods (It Fights Herpes…and Maybe Even Cancer). [online] Available at: https://draxe.com/l-lysine-benefits/ [Accessed 29 Jul. 2017].
4. Eksmatrade.lt. (2011). Amino acids are used for a variety of applications in industry, but their main use is as additives to animal feed. [online] Available at: http://www.eksmatrade.lt/amino-acids/ [Accessed 29 Jul. 2017].
5. Fda.gov. (2015). Select Committee on GRAS Substances (SCOGS) Opinion: Erythorbic acid (D-isoascorbic acid). [online] Available at: https://www.fda.gov/Food/IngredientsPackagingLabeling/GRAS/

SCOGS/ucm261286.htm [Accessed 29 Jul. 2017].

6. Hazan, R., Levine, A. and Abeliovich, H. (2004). Benzoic Acid, a Weak Organic Acid Food Preservative, Exerts Specific Effects on Intracellular Membrane Trafficking Pathways in Saccharomyces cerevisiae. Applied and Environmental Microbiology, 70(8), pp. 4449-4457.
7. One Green Planet. (2017). Need Protein? Here are 9 Amino Acids Found Abundantly in Plants. [online] Available at: http://www.onegreenplanet.org/natural-health/need-protein-amino-acids-found-abundantly-in-plants/ [Accessed 29 Jul. 2017].
8. Pappenberger, G. and Hohmann, H. (2013). Industrial Production of l-AscorbicAcid(VitaminC)andd-IsoascorbicAcid.[online]Available at: https://link.springer.com/chapter/10.1007%2F10_2013_243 [Accessed 29 Jul. 2017].
9. Rudy Mawer, C. (2016). What Is Taurine? Benefits, Side Effects and More. [online] Authority Nutrition. Available at: https://authoritynutrition.com/what-is-taurine/ [Accessed 29 Jul. 2017].
10. Webmd.com. (2017). L-TRYPTOPHAN: Uses, Side Effects, Interactions and Warnings - WebMD. [online] Available at: http://www.webmd.com/vitamins-supplements/ingredientmono-326-l-tryptophan.aspx?activeingredientid=326&activeingredientname=l-tryptophan [Accessed 29 Jul. 2017].

CHAPTER 6

FOOD COLORANTS

"When health is absent, wisdom cannot reveal itself, art cannot manifest, strength cannot fight, wealth becomes useless, and intelligence cannot be applied."

— Herophilus

INTRODUCTION TO COLORANTS

Color is vital for categorizing foods, judging their quality, and prompting aesthetic pleasure in our happenstances with them. Sincercolor is typically the primary food attribute to strike the minds, it is very significant in food marketing. All food providers including grocers, food processors etc., do their best to present a food with a strikingrcolor. In certain illustrations, the originalrcolor of the food is preserved, example thercolor of most fruits and vegetables. In other illustrations, culinary art is essential to generate new, attractivercolors, as when turkey is roasted, bread is baked, or potato chips are fried. Thercoloring matters are generally discussed under three headlines: natural foodrcolors, food browning, and foodrcolor additives.

Foodrcolors are predominant in our day to day life and are present almost in all kinds of foods and drinks. They're used to enhance the appearance of foodstuffs, for instance, they make the orangercolor of oranges more brighter and even, blue dye to marshmallows make them whiter in appearance, and much more. Other than that,rcolor dyes are used for decoration of cupcakes, candy, and multiple other food products. Foodrcolors today are used in almost all varieties of foods one can even think of. Out of the variousrcolors added to foods and drinks some of them don't have any effect on our health while other's may or may not have potential effects on human health, may be directly or indirectly. Therefore, before adding anyrcolor to any food the manufacturer needs proper approval from FDA for thercolor they are adding to the food to enhance its appearance or quality. Once they get the approval from FDA, the manufacturer also needs to mention on the back of the product with a label, if the food contains any natural and artificialrcolorant in it, with the suitable intake amount that would not cause any adverse reaction to health.

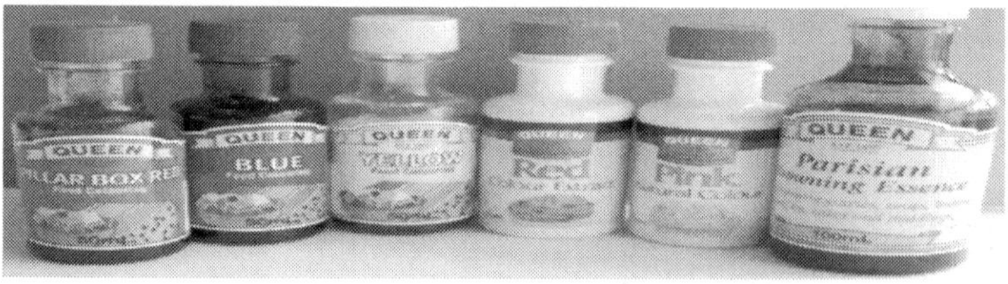

Figure 6.1: Some of the colorants added to the food products

Source: https://thankheavens.com.au/2013/12/16/how-to-know-safe-from-un-safe-food-colourcoloring/

History of food colourcolor additives

Food colourcolors in the ancient world

Naturourcolors usually obtained from vegetable and mineral sources were used ourcolor foods, drugs and cosmetics in prehistoric times. For example, the early Egyptians used synthetic or artificlourcolors in cosmetics and hair dyes, and saffron is declared as a colorant in Homer's Iliad. In ancient times tlourcolors were added to serve as a visual sign for quality, to make the products look more attractive and meet consumer expectations. Still, the use of flourcolors has also been an on-going trick for eras. There is evidence of initial pieces of regulation regarding the flourcoloring, with a 1396 French act against lourcoloring of butter, and a 1574 French law which made illegal to lourcolor to pastries.

Flourcolors in the industrial phase

During the Industrial Phase, lourcoloring of sweets with toxic chemicals was a common process, so that the sweets look more tempting to children and to adults too. In 1820, Friedrich Accum documented swelourcolored with red vermillion (mercury sulphide), red lead, white lead, yellow lead chromate, chemical mixture of copper salts of acetate, carbonate, chloride, formate, hydroxide hatesulfate, blue vitriol (cophatesulfate also known as bluestone) and Scheele's green (copper arsenite).

Until the mid of the nineteenth century, lourcolors used in cosmetics, textiles, drugs and foods were of natural source from animals, plants and minerals. In 1856, accidental discovery of the primary synthelourcolor, mauvine, was detected by Sir William Henry Perkin, who was trying to form an anti-malaria drug. A nolourcolor industry was developed, and magenta, fuchsia, violet, and a variety of blue and grlourcolors swiftly followed. Thlourcolors used in the nolourcolor industry were also used in many other industries, predominantly the textile industry. For the rising food industry, the discovery of synthetic dyes could not have been opportune. Perkins' new synthetic dyes proved to be an inexpensive and more constant techniquelourcoloring food. lourcolors demonstrated very popular and rapidly found their way into a varied diversity of foods. Numerous difficulties arose uslourcolors over-enthusiastically by the new food industry. Synthelourcolors were not only being used to mask deprived quality but permissible for fake foods to be sold as the real thing, eventually deceiving the consumer. Of concern was tlourcolors known to be toxic were being used in foods. For instance, Cheese lourcolored with red lead, sweets with mineral salts, and pickles with cophatesulfate. 'Tea' was occasionally thorn leaves dyed with copper oxide. Dr Arthur Hill Hassall gave the following grim account of the state of English food in the early 1850s:

"From morning to night (the Englishman) is the subject of perpetual fraud… he drinks chicory and beans in his coffee, water in his milk, alum in his bread, disgusting parasites, flour and gypsum in his sugar, meal in his mustard, turmeric in his gingphursulfuric acid in his vinegar, lead in his cayenne, copper in his pickles."

As the consequence, food poisoning was common and in 1851, roughly 200 people were infected in Britain, 17 of them critically, by eatlourcolored lozenges.

Imageries like these finally exploded the move for government guideline. The committee of inquiry was formed by the British Parliament to inspect Hassall's reports and this lead to the Adulteration of Food and Drink Act of 1860, which was the first act like that not only in the UK, but also in Europe. In 1874, the Society of Public Analysts was originated with Hassall as its foremost president and a committee was set up to scrutinize the working of the 1872 Act. Future amendments resulted in additional enhancements including the controllourcolorings and the choice of public analysts in all the counties and areas of Britain.

By the beginning of the 20th century, the industrlourcolor additives that are not monitored carefully had spread throughout the Europe and United States all sorts of popular foods, including ketchup, mustard, jellies and wine. The US was slower than Europe to announce national regulation. "Ketchup" was one of the foremost successfully made processed foods, with over more than 100 diverse brands of the sauce in the 1900s. ketchups are made from fermented tomato cores and skins, producers added vinegarvourflavor and preserve it, and colorant to make it red. Since the resultant sauce was prone to burst, canners started adding benzoate of soda as a preservative to it. Henry Heinz demonstrated that ketchup can be made without benzoate in a clean factory by using ripen tomatoes.

Food labels infrequently stated more than the designation and creator of the product in the early 1900s. Deprived of a regulating body, the food industry could use any element it pickedlourcolor, disguise or stretch the freshness of products. In 1906, the Pure Food and Drugs Act excluded synthelourcolors that evidenced to be "injurious to health" and the management employed chemist Dr Bernard Hesse to examine the 80 existing dyes that are used in foods and are not considered as safe to consume at the levels on which they were used. Hesse concluded that, out of the 80 or so synthelourcolors existing at the time, only 16 were more or less inoffensive and later he suggested that out of those 16, only seven of them can be used in food. The subsequent pair of times saw a course of eliminatlourcolors that instigated recurring hostile health effects in the public.

Flourcolors in the modern age of regulation

It became clear during the 1920s and 1930s, that the Pure Food and Drugs Act of 1906 did not go far beyond to guard the American public from distorted products. Accountability for imposing the Act was given to the Food and Drugs Administration (FDA) which had been recognized in 1927. In 1950, countless kids became ill by eating popcorn that contains 1-2% FD&C Orange #1lourcolor additive which got the approval to use in foods then). The FDA launched a new round of toxicological investigations. lourColor Additive Amendments of 1960 definlourcolor additive" and mentioned that only thlourcolor additives listed as "suitable and safe" for a given use can be used in foods, drugs, cosmetics and medical devices. After that for the next few years, the FDA found that numerous synthetic flourcolors caused serious hostile effects and continued to dismiss their listings until today only a minority remain authorised for use in the US.

The similar trend for the reduction in the amount of synthelourcolors used in foods can be seen in the European Union, which commenced with the launch of the Public Analysts in the 19th century, and is still important today for the safety of consumer health. The foremost European Union directive focussing on the use of colorants in foods was printed in 1962; according to whichlourcolors were allowed – out of which 20 of them were natulourcolors and rest 16 were synthelourcolors and all were measured safe for human consumption. In 1994, earlier regulation was brought up-to-date, which saw the termination of selourcolors. In 2007, Red 2G was removed due to safety concern. In 2008, new regulation cancels all previous legislation and specifies the list of permitted food additives, the foods to which they can be added and their conditions of use.

NATURAL AND ARTIFICIAL FOOD COLORANTS

Food colorings generally fall under two categories: natural food color and artificial or synthetic food color. Undoubtedly, natural food colors are the healthier substitute as they are solely derived from plants. Natural coloring typically looks less exciting when compared to artificial coloring. But this surely does not affect the taste of food.

Nacolourcolors

Naturalcolourcolors are basically pigments, dyes or other component that are obtained from vegetable, mineral or animal source that is capabcolourcoloring foods or drugs. colourcolors come from various sources such as seeds, fruits, vegetables, algae and insect. Turmeric, grass, beet roots are some of the natural sources from where colourcolors are extracted.

Some of the nacolourcolors and their derived sources are mentioned below:

- Red, violet andcolourcolor: colourcolors are derived from anthocyanins that is found in beetroots, raspberries and red cabbages.
- colourcolor:colourcolor is derived from chlorophylls, the green pigment that is found in all leaves and stems.
- Yellow, Red and ocolourcolor: colourcolors are derived from carotenoids that is found in apricots, carrots and tomatoes.

Some of the natural colorants used in food are-Carotenoids, chlorophyllin, anthocyanins, and betanin they cover four chief categories of plant pigments grocolourcolor food products. Other colorants or dedicated derivatives of these essential groups include: **Annatto**, which is a reddish-orange dye made from the seeds of the achiote; Cacolourcoloring, that is made from caramelized sugar; Carmine, which is a red dye derived from the cochineal insect, Dactylopius coccus; Elderberry juice; Lycopene; Paprika; and Turmeric.

Orange or yellow pigments. Think beta-carotene or the orange color of carrots. Carotenoids are also not used too often in food products due to their sensitivity to oxidation, which can cause hue change.

Carotenoids-

Some of the most common carotenoids are lycopene and beta-carotene (precursor of vitamin A). Xanthophyll lutein in plants, is the most plentiful carotenoid and its role in averting age-related eye disease is presently under examination. Because of the masking presence of the chlorophyll, lutein and the added carotenoid pigments that are found in mature leaves are habitually not noticeable. In the absence of chlorophyll, for example, as in autumn foliage, the carotenoids yellow and colourcolor are predominant. Due to the presence of carotenoids characte colourcolor to carrots, daffodils, corns, egg yolks, buttercups, bananas etc. are given.

Carotenoids are mainly responsible for the bright yellows and oranges hue of deciduous foliage of certain hardwood species like hickories, cottonwood, aspen, black cherry, alder, birch, sycamore, maple, ash, yellow poplar, and sassafras. Carotenoids are the leading pigment in autumn leaf coloration.

Figure 6.2: *Some of the carotene rich foods*

Source: http://www.faceonclinic.com/health-tip/carotenoids-for-uv-skin-protection/

Anthocyanins-

These are the red, purple, e colourcolor pigments that are found in the leaves, stems, roots, flowers, and fruits of a. ColourColor and concentration of these compounds can be altered simply with pH, which permits for ns colourcolor potentials. General sources for anthocyanin compounds comprise grape skins and beet roots.

The six anthocyanins that are used d colourcoloring are found in the following fruits and vegetables mentioned below:

- Cyanidin –Blackcurrants, Raspberries, elderberries, red cabbage
- Delphinidin – blackcurrants, Blueberries
- Malvidin – Grape
- Pelargonidin - Strawberries and radishes
- Peonidin – Cranberries
- Petunidin – Blueberries

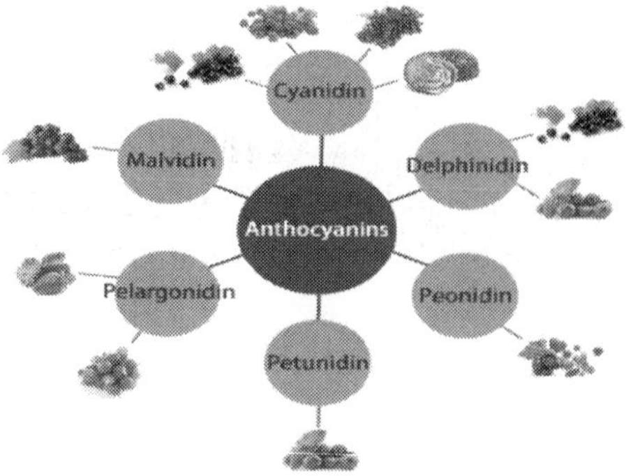

Figure 6.3: Anthocyanins useood colourcoloring

Source: http://jtbpd.com/information.php?information_id=41&xcSID=obgk6g flk6kcqbfqr39qr03u32

Anthocyanins are water soluble and are suitable for low tems colourcoloring. For most of the anthocyanins, for obtaining intense able colourcolors, the pH should be lower. With the increase in pH, the intensity of anthocyanins loses and they become bluer. At the pH, higher than 4.5 they become like blackcurrent pigment.

Figure 6.4: Some of the anthocyanin rich food

Source: https://agroekonomija.wordpress.com/tag/organska-hrana/

Anthoxanthins-

These a white, colourcolorless, off-white, or yellow pigments that occur in most plants. Anthoxanthins don't occur too repeatedly in food products as additives, but they occur in many vegetables and fruits.

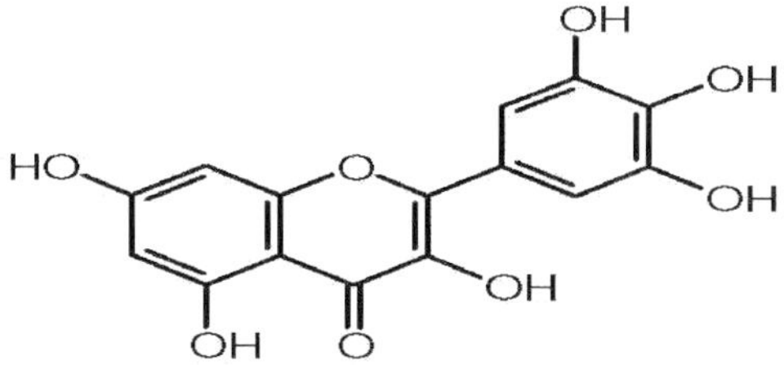

Figure 6.5: *Structure of anthoxanthin*

Source: https://alchetron.com/Anthoxanthin-1719486-W

Chlorophylls-

These are bright togreen in colourcolor. Chlorophyll as the substance in plants that let the plant to absorb light for photosynthesis, as well as they also are responsible for making many plants green.

Chlorophyll are not found as colorant in many foods dur to the reason that are very instable, so they are only found in vegetables.

Figure 6.6: Chlorophyll rich foods

Source: http://www.healthline.com/health/chlorophyll-bad-breath#chlorophyll-facts1

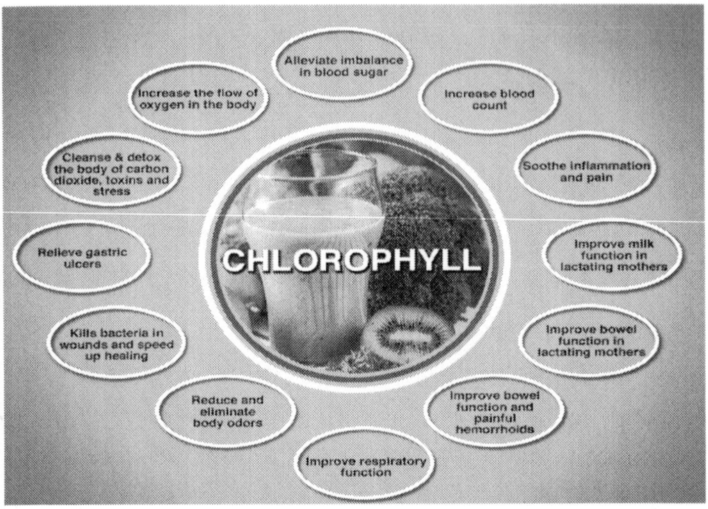

Figure 6.7: Above picture depicts the advantages of chlorophyll

Source: http://drjockers.com/the-power-of-green-foods/

Annatto

Annatto is ane-red food colourcolor resultant of the seeds of the *Bixa orellana* (achiote tree). It is habitually used to give or orange colourcolor to foods, sometimes annatto is alsto impart flavourflavor and aro foods. Its odourodor is "somewhat peppery with a tint of nutmhe annatto colourcolor is derived from several carotenoid pigments, mostly bixin and norbixin that are originated from the reddish waxy coating of the seeds. Annatto and its extracts are commonly used in an indusscale as a colourcoloring agent in various processed food products, like- cheese, butter, cakes, smoked fish, breakfast cereals, sausages, and many more. Annatto is a considered as a natural substitute toicial food colourcoloring compounds, but use of annatto in excess amount also leads to various food-related allergies. Annatto is of specific marketable value in the United States because the Food and Drug Administration declares that colorants derived from annatto to be "exempt of certification". Since 16th century, Annatto is used as an ancient colorant for Gloucester cheese.

Annatto is presently used to give or orange colourcolor to countless industrialized and semi-industrialized foods. It is recognized by the E number E160b, by the European Union.

Figure 6.8: The annatto seed

Source: http://bilgiliyem.com/bitkiler/anaatto-nedir-nerelerde-yetisir-faydalari-nr/

Caramel ColourColoring

It is a woluble food colourcoloring. It is prepared by heat treatment of carbohydrates, in presence of acids, alkalis, or salts, the process called as caramelin. It has an odourodor of burnt sugar and a slightlyr taste. Its colourcolor varies from pale yellow to orange to down. Caramel colourcoloring is one of the most initial andly used food colourcolorings that is found in numerous foods and beverages produced commercially. Some of them are- beer, brown bread, buns, chocolate, cookies, cough drops, spirits and liquor such as brandy, rum, anky etc. Caramel colourcoloring is extensively permitted to be used in food worldwide but its application and usage varies from country to country.

Carmine

Carmines are also referred to as cochineal, cochineal extract, crimson or carmine lake, natural red. It is aof a bright-red colourcolor which is found in the aluminium salt of carminic acid. This tint is formed from few scale insects like- the cochineal scale and certain Porphyrophora species. Carmines are used in the manufacturing of synthetic flowers, crimson ink, paint, rouge and other cosmetics. It is also used in the manufacturing of certain medications. It is regularly mixed with the food products like- yogurt, candy and certain brands of juice, etc.

According to the United States of Food and Drug Administration (FDA), in the year 2006, it was proposed that the food products that contain carmine needs to be listed on the ingredient label by their name. Apart from this, it was also reported that the FDA will independently analyze the ingredient labels of thcines that carry colourcolorings that is procured from carmine. The Centre for Science in Public Interest claimed FDA to state on ingredient label that carmine may cause serious allergic reactions and anaphylactic shock, etc. Although there were many concerns regarding the use of carmine as a colorant as it causes allergic reactions, but FDA has not barred carmine and declared that carmine has no significant hazard risk to general population.

The European commission directives regulates the use of carmine in the European Union. The directive governing food colorants supports the use of carmine only for few sets of foods. Even though use of carmine in foods raise the concern of hazard, it is yet not banned by the European Union. Nevertheless, the European Food Safety Authority (EFSA) has demoralized the used of carmine in foods. Carare still used as colourcolorings in alcoholic beverages predominantly.

Figure 6.9: Carmine uses in different foods

Source: https://www.pinterest.com/pin/551479916841813073/

Paprika Oleoresin

Paprika oleoresin or paprika extracts are the oil-soluble extracts that are gained from the fruits of *Capsicum annuum* or *Capsicum frutescens*, this icipally used colourcolors and flavourflavors to the food products.

Paprika extracts are composed of cap which is the core flavourflavoring compound giving sharp smell or taste in higher concentrations, as well as capsanthin andrubin, are the core colourcoloring compounds.

Figure 6.10: Paprika oleoresin obtained from capsicum annuum

Source: http://www.lalaessentialoils.com/paprika-oleoresins-40000-cu.html

Paprika Oleoresin extracts are used in processed meat, popcorn oil & cheeses, beverages, snack food coatings, and tomato products. Since it is an oil soluble produ is used in many food colourcoloring applications. Various other applications where paprika oleoresin is used are- cheese, orange juice, spice mixtures, poultry for attaining deepened colourcolor oyolk.

Syntheticcolourcolors

Synthetic colourcolorslso known as artificial colourcolors. They are manufactured by chemical reaction and are frequently used in food and pharmaceutical industries. Some of the most commonetic or artificial food colourcolors are tartrazine, sunset yellow, amaranth, Allura red, quinoline yellow, brilliant blue and indigo carmine.

Tartrazine

Tartrazine is also referred to as yellow dye #5. Tartrazine is used to impart bright lemon-yellow color to foods. According to some reports, tartrazine is the second most frequently used dye after Allura red. Tartrazine is also called as an azo dye and the chemical formula of tartrazine is $C_{16}H_9N_4Na_3O_9S_2$. The chemical formula of tartrazine also contains few other elements other than nitrogen, hydrogen and, these other elements are SulphurSulfur, sodium and oxygen. Azo (derived from French word meaning, nitrogen) dye is a chemical compound which combines two hydrocarbon groups with the two nitrogen atoms. Azo dyes reports for approximately 60 to 70 percent of all dyes that are used in the food industry as well as in textile industry. Azo dyes are very cheap to produce and are much more stable than natural food dyes, this is the reason behind their so much popularity.

Figure 6.11: Tartrazine dye used as colorant in nacho chips

Source: http://idsgn.org/posts/an-edible-color-palette/

Tartrazine dyes are used in broad range of food products as colorants including desserts and candies, condiments, soft-drinks and breakfast cereals. Besides this tartrazine is also used in cough syrups and in cosmetics too. Below is the list of few foods that contains tartrazine-

- Canned Pie Filling
- M&M's Candy
- Kraft Macaroni and Cheese Dinner
- Mountain Dew
- Cap'n Crunch Cereal

Tartrazines dyes may also cause intolerance in people who suffer from asthma or aspirin intolerance. The tartrazine reactions range from mild to severe depending upon person to person and their tolerance level. Tartrazines often leads to hypersensitive reactions, when combined with sodium benzoate, it leads to hypersensitivity and other behavior changes in children. The E- number of Tartrazine is 102.

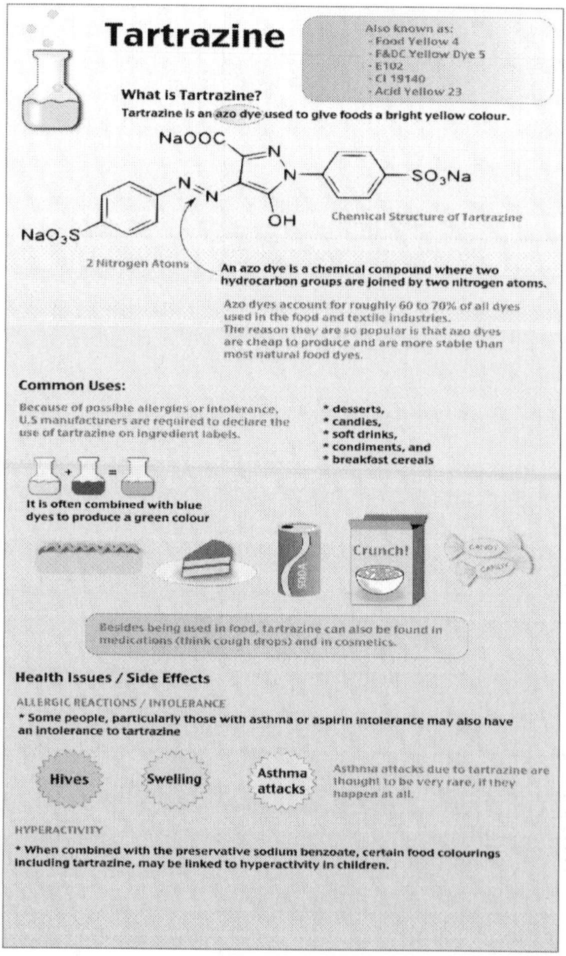

Figure 6.12: The above picture depicts the complete role of tartrazine, its uses, and its side effects

Source: http://foodconstrued.com/wp-content/uploads/2012/09/Tartrazine.png?x71829

Sunset Yellow

Sunset yellow is also referred to as yellow dye #6. Sunset yellow imparts orange-yellow color to foods. Analogous to tartrazine, sunset yellow is also an azo dye. The chemical formula of sunset yellow is $C_{16}H_{10}N_2Na_2O_7S_2$. Sunset Yellow is used in various diverse kinds of products, ranging from soft drinks to candies and snack foods. Also used in coloring of dessert powders, gelatin deserts, sausage, drugs and cosmetics. Some of the specific products that fall under this category are:

- Orange Crush
- Doritos
- M&Ms Candy

Sunset yellow often leads to allergic reactions and intolerance particularly to those who already show sensitivity towards aspirin. Sunset yellow also cause adrenal tumors in animals and serious sensitivity reactions. The E-number for Sunset Yellow is 110.

Figure 6.13: The sunset yellow dye used as an artificial colorant in juices and drinks

Sources: http://idsgn.org/posts/an-edible-color-palette/

Amaranth

Amaranth is also referred to as red dye #2. Amaranth dye, gets its name fe plant amaranth which has the similar colourcolor like this dye has. Amaranth is also an azo dye like tartrazine and sunset yellow. Amaranth was barred in the United States in the year 1974 because it was suspected to be carcinogenic and as a substitute Allura red was used in its place. Though its usage was continued to other countries, counting- Canada. Originally, the amaranth tint was prepared from coal tar, but currently most of the synthetic dyes are made from petroleum by-products. Amaranth is used as a food colorant in alcoholic beverages, cake mixes, ice cream, red soft drinks, jellies, jams, and numerous other processed foods. In France and Italy, amaranth is used in the production of caviar. Amaranth dye cause severe reaction to those who show salicylate intolerance. Furthermore, amaranth is also a histamine liberator, and hence intensifies the signs of asthma. The E-Number of amaranth is 123.

Figure 6.14: Amaranth "hopi red dye"

Source: http://www.dianeseeds.com/amaranth-hopi-red-dye.html

Figure 6.15: Ice-creams colored with amaranth dyes

Source: https://www.hawkinswatts.com/ingredient/amaranth/

Allura Red

Allura red, also referred as red dye 40ra red is an artificial dye that provides colourcolor to foods, cosmetics, and medications. Allura red is produced by the by-products of the petroleum industry. Just like the above three, Allura red is also an azo dye. Allura red as the name says looks like a dark red powder, and is the most widely used and consumed dye. Many European countries like- Denmark, France, Belgium, and Switzerland prohibit the use of Allura red as a colorant in food. The Centre for Science in the Public Interest, in the United States has been demanding FDA for prohibition of Allura Red.

Allura red is generally used in processed foods such as baked goods, candy, cereal, dairy products, drinks, sauces, and snacks. Although, Allura red is particularly a non-toxic dye, but people who have salicylate intolerance also shows similar reactions towards Allura red. Moreover, it is also a histamine liberator, and it also intensifies the symptoms of asthma. One of its degradation products leads to bladder cancer in animals when existing in high concentrations. The E-number of Allura red is E 129.

Figure 6.16: The Allura red dye made from petroleum

Source: http://www.foodadditivesworld.com/articles/red-food-coloring.html

Ponceau 6ceau 6R is a bright red dye that is used to colourcolor food. The name of this dye comes from the French word "poppy". It belongs to the family of dyes named Ponceau dyes. Other names that are found in this class are- crystal scarlet, acid red 44, or C.I. 16250. Ponceau dyes are also azo dyes with the chemical formula $C_{20}H_{12}N_2Na_2O_7S_2$. It is not a widely used dy dye is only found in jelly crystals, fruit-flavourflavored fillings, and cake mixes.

This dye is permitted both in United States and European Union. Apart from its use as a colorant, this dye is also used for studying cells. It is used as a stain to see the structures under the microscope clearly. This dye didn't show any adverse effects except that there is possibility that this dye often leads to sensitive reactions in people who are allergic to aspirin and benzoates. The E number of Ponceau 6R is 126.

Figure 6.17: Porceau 4R "the bright red dye"

Source: http://www.aciddyesmanufacturer.com/ponceau-4R-Food-Colors.htm

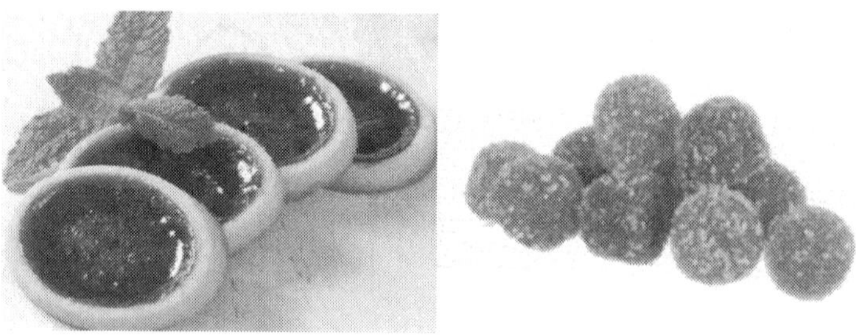

Figure.18: Porceau 4R dye found in jellies and fruit-flavourflavored fillings

Source: https://www.hawkinswatts.com/ingredient/ponceau-4r/

Carmoisine

Carmoisine is a red to maroon color dye. It is in an azo dye. Carmoisine is also known as Azorubine, Food Red 3, Azorubin S, Brillantcarmoisin O, Acid Red 14, or C.I. 14720. Carmoisine is frequently used when heat-treatment is given to foods. Carmoisine is not approved to be used in the United States and Canada, whereas in Europe, the use of this dye is continued. The E-Number of Carmoisine is 122.

Some of the common foods in which this dye is used includes-

- Baked Products
- Condiments
- Cakes
- Candy and Cough Drops
- Drinks
- Ice Cream
- Jelly Crystals.

Figure 6.19: Carmoisine dye used as colorant in cakes and drinks

Source: http://foodcolors.org/food-colors/carmoisine-food-colors/

Erythrosine

Erythrosine is also referred to as red dye #3. It is a cherry-pink food colorant that is made from coal tar. It is an organic compound containing iodine and sodium. Erythrosine is also called as xanthene dye. Xanthene dyes are called by this name because they contain a xanthene molecule at their base and their color ranges from yellow to pink to bluish red in color. The molecular formula for xanthene is $C_{13}H_{10}0$.

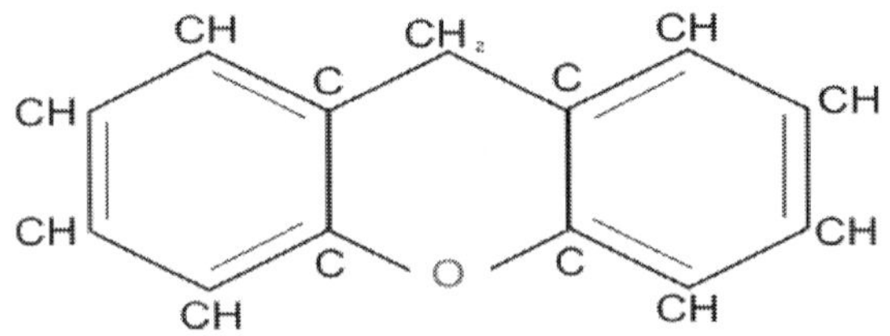

Figure 6.21: Xanthene molecule

Source: http://foodconstrued.com/2012/05/erythrosine/

The chemical formula for Erythrosine is $C_{20}H_8I_4O_5$.

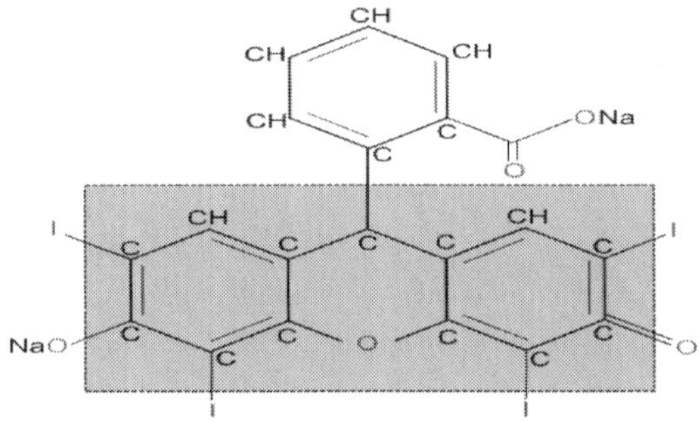

Figure 6.22: Erythrosine molecule

Source: http://foodconstrued.com/2012/05/erythrosine/

Erythrosine is principally used as a food dye. Some of the most common uses of erythrosine include: cocktail and candied cherries (Maraschino cher candies, popsicles, cake decorating gels, used to colourcolor pistachio shells. In United State, erythrosine is used very frequently as in place of this Allura red is preferred more. In 2008, the Centre for Science in the Public Interest appealed the FDA for a complete prohibition on erythrosine but not actions have been taken by FDA so far. The United states citizens can still use this dye without any restriction from FDA.

The E-number of erythrosine is E127. Erythrosine was recognized in 1990 by the FDA as a thyroid carcinogen in animals and is banned in cosmetics and externally applied drugs. Used in sausage casings, oral medication, maraschino cherraked goods and candies.

Figure 6.20: Pistachios colourcolored with erythrosine dye

Source: http://idsgn.org/posts/an-edible-color-palette/

Citrus Red

Citric acid is toxic to rodents at modest levels and caused tumors of the urinary bladder and possibly other organs. Since 1956, the use of citrus red has been permitted by the United States, food and drug administr(FDA). This food dye is used on skins of oranges to colourcolor them. This food dye is only permitted to be used on the peel of oranges but this dye didn't get any approval to be used in manufactured orange juices. This dye is listed as group 2B carcinogen by the International Agency for Research on Cancer (IARC). Group 2B carcinogens are "potentially carcinogenic to humans". In United States, citrus red is freely used in the states of Florida, but prohibited in California and Arizona.

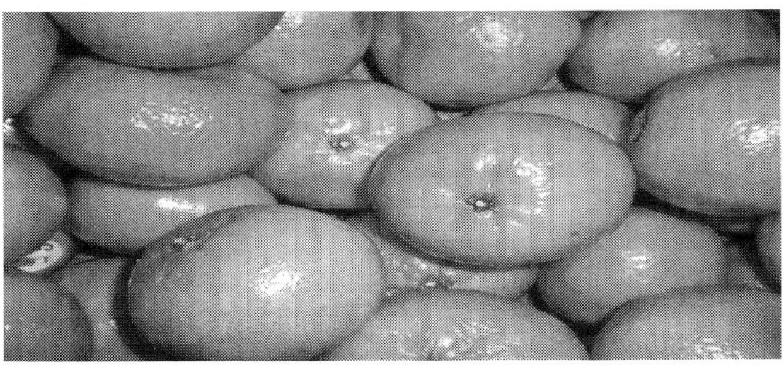

Figure 6.21: Skin of oranges dyed with citrus red

Source: http://thescientistgardener.blogspot.com/2010/02/orange-dyed-oranges.html

Fast Green

Fast green dye is a sea-green triaryl methane food dye. The use of this dye is banned in the European Union and several other countries. In United States, fast green dye is used least in foods, but not yet banned. Fast greens are used as a colorant for tinned green peas and other vegetables, fish, desserts, etc.

Figure 6.22: Green peas dyed using fast green dye

Source: http://idsgn.org/posts/an-edible-color-palette/

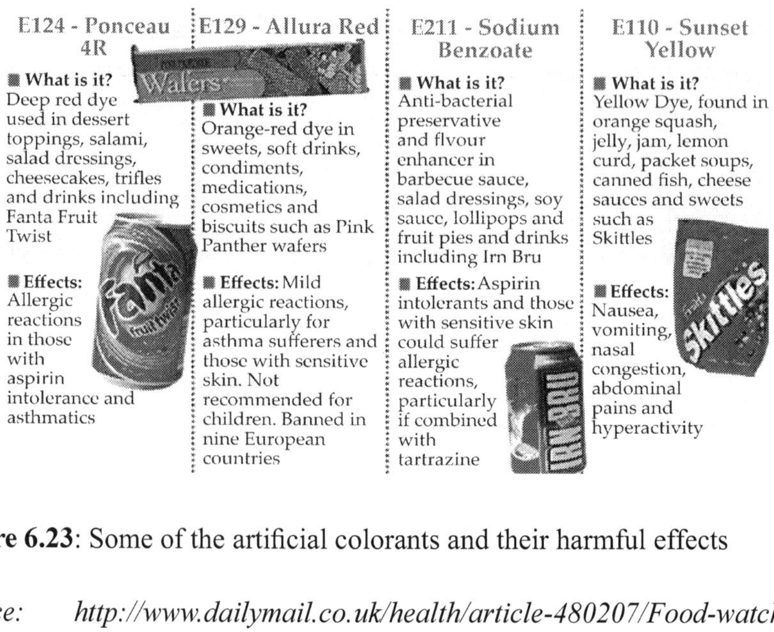

Figure 6.23: Some of the artificial colorants and their harmful effects

Source: http://www.dailymail.co.uk/health/article-480207/Food-watchdog-condemned-totally-inadequate-response-harmful-food-additives.html

Synthetic or Natural

Talking about food colorants or any food additive people always prefer the natural ones over any synthetic option. What consumers may not be aware about products labelled "natural" is that there is no regulated definition for them that exists. Natural is tough to express and has only begun to matter to consumers. Different people have different definition to the term "natural". Although FDA and the other regulatory bodies around do not have a definition for natural, food companies like Whole Foods set their own values for what they consider to be natural and what are their requirements to include in thects. The FDA controls all food colorants and additives. ColourColors fall under one of two types, certified and exempt from certification. Certified colors are those that are produced synthetically, used extensively, and are very less expensive. Nine of these colors have got approval for use in the U.S., these are: blue 1, blue 2, green 3, red 3, red 40, yellow 5, yellow 6, citrus red 2, and orange B. Colors that don't need any certification are those derived from natural sources like vegetables or minerals. They are typicall expensive, less stable, and frequently add accidental flavourflavors to foods. Both groups of colorants are verified for safety and must be used conferring to certain rules and quantities.

The labelling of foods holding colorants might be puzzling sometimes. To sum it up all there are three guidelines that needs to be kept in mind:

Natural colors do not exist when it comes to food, only natural sources from where they have derived exists.

A food that comprises "artificial colors" in its ingredient label does not certainly mean that the colors used in the food are synthetic. Synthetic colors, like blue 3, must be registered in gredients list. "Artificial colors" only means that the colourcolor that is added to the food is not inherent to it. For instance, beet root juice is o give strawberry ice cream and yohe typical pink colourcolor which is the characteristic colourcolor of strawberry and it is measured artificial because beets aren't typically included in strawberry ice cream or yogurt recipe.

When precisely seeing for products that use colorants from natural sources, always look for labels that says, "no artificial colors used" or "colors from natural ingredients/sources".

Colors, make the food more enjoyable both in terms of preparation and eating, no matter what source they come from. Whether it is natural or synthetic, the important objective is to meet the anticipated conditions of the product as specified by the regulation. There are numerous regulations set by the FSSAI-the Indian regulatory body for the use of food colorants in diverse foods. Even if the slope of natural product tils and it does not meet theed specifications, there is no purpose of using natural colourcolors too.

REFERENCES

1. All the Colors of The Rainbow: Food Colorants. (2015). [Blog] kulikulifoods.com. Available at: http://blog.kulikulifoods.com/2015/03/19/all-the-colors-of-the-rainbow-food-colorants/ [Accessed 29 Jul. 2017].
2. DiLeo, M. (2010). Orange dyed oranges. [Blog] Thescientistgardner.com. Available at: http://thescientistgardener.blogspot.in/2010/02/orange-dyed-oranges.html [Accessed 29 Jul. 2017].
3. Fda.gov. (2016). Color Additives. [online] Available at: https://www.fda.gov/forindustry/coloradditives/ [Accessed 29 Jul. 2017].
4. Food - Construed. (2012). Tartrazine: What is it and why is it in my food? [online] Available at: http://foodconstrued.com/2012/09/tartrazine/ [Accessed 29 Jul. 2017].
5. Safefood.eu. Food Colour Resource | The history of food colour additives. [online] Available at: http://www.safefood.eu/Food-Colour-Resource/History.aspx [Accessed 29 Jul. 2017].
6. The Gluten Free Lifesaver. How to know Safe from Unsafe Food Colouring. [online] Available at: https://thankheavens.com.au/2013/12/16/how-to-know-safe-from-unsafe-food-colouring/ [Accessed 29 Jul. 2017].
7. What-when-how.com. INTRODUCTION (Food Colors). [online] Available at: http://what-when-how.com/food-colors/introduction-food-colors/ [Accessed 29 Jul. 2017].

CHAPTER 7

ENZYMES AS FOOD AND FEED ADDITIVES

"Any food that requires enhancing by the use of chemical substances should in no way be considered a food."

—John H. Tobe

Enzymes are the biological catalysts. Enzyme preparations obtained from the extracts of plants or animals are in use much before the properties or nature of enzymes were known. Food and feed industry continuosly uses various enzymes obtained from different sources to convert the raw-material to a desired product.

In this chapter, we will discuss the historical and modern era of enzyme production. Application of major enzymes in both food (enzymes like- cellulase, amylase, glucose oxidase, pectinase, α-Acetolactate decarboxylase, etc.) and feed (enzymes like- phytase, carbohydrase, and protease) industry. Also, the concept of genetically modified Enviropig is discussed in later sections.

ENZYMES AS ADDITIVES IN BIOTECHNOLOGY

Enzymes are crucial in the metabolism of all living organisms. Enzymes are used to drive the chemical reaction with the increased rate. Particularly, the use of biocatalysts for the processing of raw materials and as additive has been in practice since long. Enzyme preparations obtained from the extracts of plants or animals are in use much before the properties or nature of enzymes were known.

Food industry is continually looking for the progressive skills to meet the requirements of the consumers, and enzymes are used as a foremost tool for the conversion of raw materials to the desired end-products by the industrial product makers. After getting clean certification from GRAS, the enzymes use in food biotechnology is driven extensively. When the purified enzymes are added to food products or food preparations, they tend to enhance flavourflavor, digestibility, texture, and nutritional value. It was just 30 years back since the use of commercial enzymes has fully grown up in the food sector, after the speedy expansion in the protein technology has occurred. Enzymes have gradually become the vital aspect for the manufacturing of vegetables, meat, baked goods, fruits, milk-products, and both alcoholic and non-alcoholic beverages. Over the past 10 years, the number of articles have been increased describing the improved product yields, both in food and beverage manufacturing. Furthermore, since it is required in diverse divisions of food technology to modify the physical and chemical properties of protein, some of the earlier unexplored enzymes are presently active to produce variability of foods in which the biocatalysts substitute theoretically carcinogenic or other destructive chemicals. This also comprises new methods where the characteristics of natural products are transformed to fit the nutritional or technical requirements changing.

The monetary advantage of using technical enzyme preparations lies in dropped process costs, use of renewable resources to reduce on the environmental impact, and regularly increasing the quality of the products. Similarly, preservation also makes a noteworthy impression on the quality of foods and beverages. For instance, today the technology has grown so fast that it is possible to convert juices into concentrates, these concentrates can be stored for longer durations without losing the quality, except for aroma. Another example of improved preservation is- stabilizing flavourflavor and colourcolor. After the arrival of biotechnology and its advancements in diverse fields, it has now become possible to find the methodologies that suggests solutions to the persistent problems and opportunities to open new possibilities. Amongst these advancements, enzymes are projected as model agents of "green" technology since they are used to treat the biological wastes or to avert their development. Presently the enzymes that are used are originated from either plants or animals, but most of them are originated from the beneficial microorganisms. Consequently, many purified enzymes are now being extensively used in food additives, other than in food processing. It is important to note here that like all proteins, enzymes can also cause reactions to the body if the person is exposed to larger quantity of them. Therefore, they should be used in adequate amounts to prevent allergic reactions or any other harmful effects.

After the innovation of this particular subject of Enzyme Research, diverse number of enzyme applications are contributing to the food technology, most of them are related to explain how different biocatalysts helps in bringing improvements in food processing.

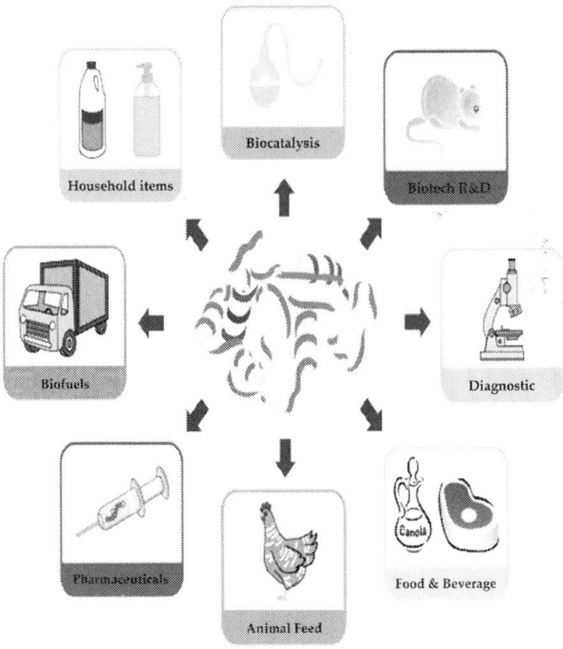

Figure 7.1: Common industrial uses of enzymes

Source: http://ib.bioninja.com.au/standard-level/topic-2-molecular-biology/25-enzymes/enzymes-in-industry.html

Enzyme history in food production

Traditionally, enzymes were extracted from the edible parts of the plants, tissues of food animals, and also produced by the microorganism's like- yeast, bacteria, fungi. An example of natural enzyme extracted from the stomach of calves and other domestic animals is rennet, this natural enzyme is in use since long in cheese making. Rennet comprises a protease enzyme which coagulates milk, instigating it to separate into solids that is curd; and liquids that is whey. Other than this another enzyme that are produced by yeast are used in the grape juice fermentation to make wine. The definition of enzymes is still the same since centuries, they are the proteins that acts as catalyst and increase the rate of chemical reaction in humans, microorganisms, plants, animals, and other living organisms. Only few enzymes can increase the rate of reaction up to ten million times.

Enzymes requires proper conditions to work, for example, an optimum acidic pH and temperature, etc. Based on the type of reaction enzymes catalysecatalyze and the substrate they act upon they are classified. For instance, lactose is catalyzed by enzyme lactase, protease is catalysedcatalyzed by enzyme protease, and lipase is catalyzed by enzyme lipase and many more. Other than this, several long-used enzymes have common names, like papain, from papaya, this enzyme is used to tenderize meat.

Modern production of food enzymes

In the twentieth century, enzymes commenced to be isolated from the living cells, leading to the large scale commercial production of enzymes with broader application of these in food industry. Presently, microorganisms are the major sources of commercial enzymes. Though the microorganisms do not cover the same enzymes as plants or animals does, it only produces a related enzyme that catalyzes the desired reaction only.

Enzyme manufacturers have augmented microorganisms for the manufacturing of enzymes via natural selection techniques and classical breeding techniques. There are several specific methods, like- direct genetic modification, by which enzymes can be produced by optimizing microorganisms. These methods help in obtaining high-yielding producing organisms. With the help of various biotechnological tools, today it is possible to transfer the microorganism that are not adequate for commercial enzyme production to a safe history of enzyme production for use in food. The biotechnological produced enzymes are alike those that occur or found in nature. Moreover, the enzymes that are produced by microorganisms are extracted and disinfected before their usage in food manufacturing.

What Do Enzymes Do for You ?

SECTOR	APPLICATION AREA	BENEFITS
• Detergents	Household washing and cleaning agents	Wash your clothes in cold water; make your teeth cleaner
• Textiles	Denim washing, silk polishing, leather goods softening	Stonewash your jeans; make cotton look and feel like silk; make your leather soft
• Food processing	Baking, brewing fruit juice processing	Clarify your juice and beer; make bread better; turn corn starch into sugar syrup
• Pulp and paper	Starch conversion, pitch control, bleach-boosting, deinking, stickies control, slime control	Reduce production costs and improve quality

Figure 7.2: Role of enzymes in different industries

Source: http://www.edt-enzymes.com/usingindustrialenzymes.html

Enzyme regulation in Canada

Enzymes used in food processing modulate as food additives in Canada. According to the section B.01.001 of Food and Drug Regulation, the enzymes encounters the definition of food additive, while affecting the characteristic of the food or when its by-products become part of it. The physical enzyme residue is considered while determining the remaining enzyme residue in food. The Canada government conducts pre-assessment test for the enzymes and assent their use in foods, just like any other food additives are evaluated. Enzymes that pass the pre-market safety assessment test and get permit to be use in foods along with the permitted source of each enzyme, they are then further listed in the List of Permitted Food Enzymes published on Health Canada's website. The source organism's safety is the primary deliberation in assessing an enzyme product. For instance, some of the food animals and the edible plants have safe history to be used as source enzymes in food industry. The microorganism that are used in food enzyme production should be well-characterized and should not produce any pathogens, antibiotics, or any other harmful toxins. Health Canada's safety assessment reflects toxicity tests on the enzyme product and the process used by them to produce the enzyme commercially.

MAJOR APPLICATIONS OF ENZYMES IN FOOD INDUSTRY

Enzymes have been used to produce or improve the quality of food in the food industry. Some of the products like- cheese, bread syrup, yogurt, etc, used variety of enzymes. Traditional methods like brewing, meat tenderization, cheese making etc. rely on proteolysis, these were developed prior to the understanding of enzymes. Early food processes concerning proteolysis were usually the unintentional consequence of endogenous or microbial enzyme activity in the foodstuff. Some of the major applications of enzymes are:

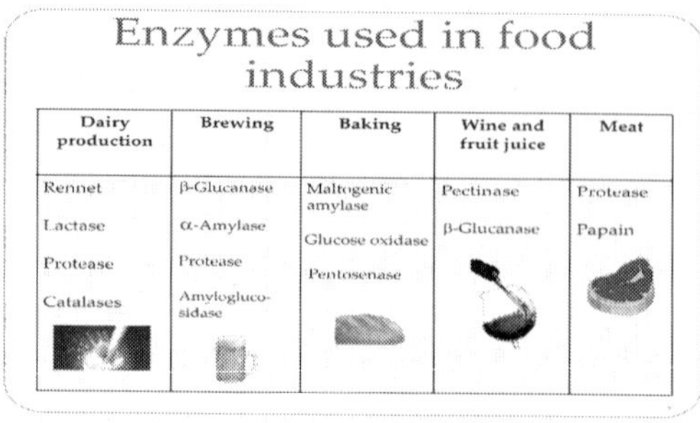

Figure 7.3: Use of enzymes in food industry

Source: https://www.slideshare.net/shweta1095/enzymes-in-food-industry

Rennet

The rennet use in the making of cheese is one of the earliest methods of exogeneous enzymes in food processing. This method of cheese manufacturing has the largest application of enzymes in food processing industry. After rennet, proteinases have gained further applications in dairy industry in the recent years. Proteinases are used in accelerating cheese ripening, in modifying the properties a preparation of dietic products. Currently, rennet is used as one the third most widely used methods for production of cheese worldwide. Rennin acts on milk protein in two steps, one by enzymatic action and other by non-enzymatic action, resulting in milk coagulation. Various microorganisms produce rennet-like protease which are the substitutes of calf rennet. Microorganisms such as- Rhizomucor pusillus, endothia, parasitica, etc. are used broadly for rennet production in cheese production. The microbial rennets are also produced by the fidderentdifferent strains of Mmucor species.

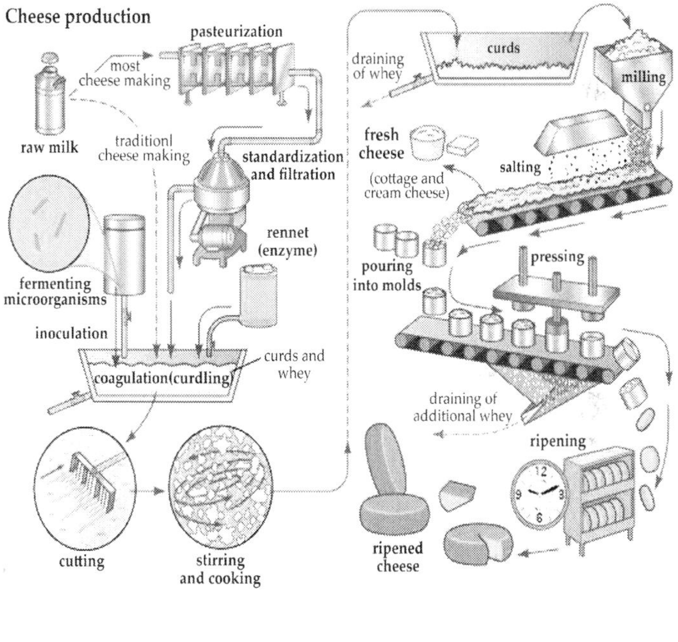

Figure 7.4: The process of cheese making using rennet enzyme

Source: https://www.britannica.com/topic/dairy-product/Cheese

Lactase

Lactose is the form of sugar that is found in milk and whey and its lactase, hydrolasehydrolase, or beta-galactosidase. The enzyme immobilization technique has given new understanding and possibilities for utilization of this sugar. But due to intestinal enzyme insufficiency some enzymes show lactose intolerance and trouble in consuming milk and dairy products containing lactose in it. Therefore, to overcome such cases, low-lactose or lactose-free food are advised for such individuals who are intolerant to lactose, to prevent them from severe tissue dehydration, diarrhoeadiarrhea, or more severe cases of intolerance often leads to death of the individual. The lactose- treated milk also has other advantages than the ones mentioned above, like- lactose-treated milk increases the sweetness of the resultant milk, and hence, avoid the need of addition of sugars in the manufacturing of flavored milk drinks. To enhance the scoop and creaminess, digestibility, and sweetness, the ice-cream manufactures uses lactases.

Lactose can be attained from several plant sources, animal sources, several microorganisms, or mouldsmolds. Some of which are used in commercial enzyme preparation. Some of the lactose preparations obtained from Aspergillus niger and, A. oryzae etc, are measured to be safe, as these sources have undergone various safety tests already. The enzymatic hydrolysis of lactose can be accomplished either by free enzymes, typically in batch fermentation process, or by immobilized enzymes or by immobilized whole cells making intracellular enzyme. Though several hydrolysis systems have been inspected, only limited of them have been scaled up with achievement and even rarer have been useful at an industrial or semi-industrial level. Numerous acid hydrolysis systems have been established to industry-scale level. Many commercially immobilized systems have been settled for commercial exploitation. In Italy, Snamprogretti process of industrial-scale milk processing technology is one of the employed system. These systems use fibrefiber-entrapped yeast lactase in batch process, and the milk used is formerly sterilized.

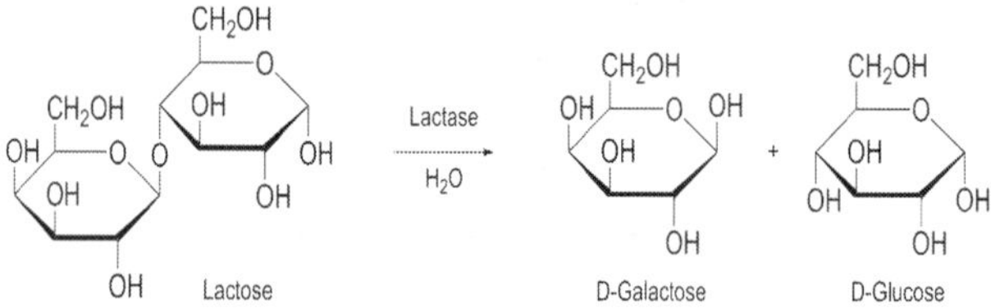

Figure 7.5: Breakdown of lactose with the help of enzyme lactase

Source: http://ib.bioninja.com.au/standard-level/topic-2-molecular-biology/25- enzymes/enzymes-in-industry.html

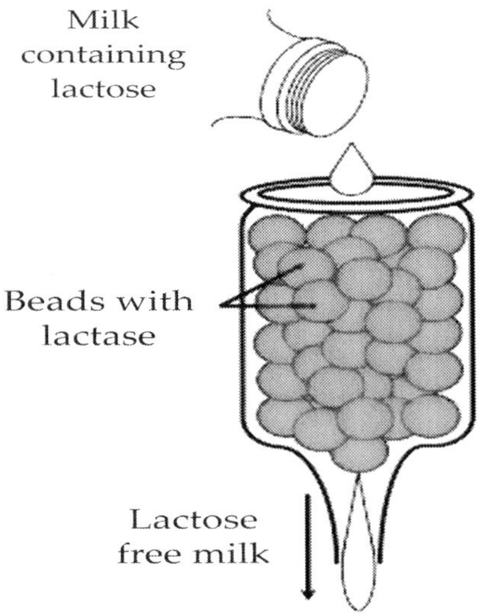

Figure 7.6: Generation of lactose-free milk using enzymes

Source: http://ib.bioninja.com.au/standard-level/topic-2-molecular-biology/25- enzymes/enzymes-in-industry.html

Catalases

Enzyme catalase is produced from bovine liver or from microbial sources. This enzyme breaks down hydrogen peroxide into water and molecular oxygen. Catalase enzymes successfully removes the residual hydrogen peroxide also, this ensures that the fabric is peroxide free and is chiefly used in food industry and in egg processing with several other enzymes. This enzyme also has a partial use in cheese production process.

Catalase is one of the most common enzymes that is found in almost all living organisms who are exposed to oxygen, and then their function is to catalyze the decompose hydrogen peroxide to water and oxygen. Both catalase and glucose oxidase are frequently used together in particular foods for preservation. Superoxide dismutase is an antioxidant that is used in foods, but acts effectively in presence of catalase.

Lipases

Lipase, which is a water-soluble enzyme helps in catalyzing the hydrolysis of ester bonds in water insoluble, lipid substrates. Lipases are either produced individually by the microorganism or with the esterase. Some of the lipase producing microorganisms are- Serratia marcescens, Pseudomonas aeruginosa, etc. Lipases act as a catalyst and they help to produce fatty acids, glycerol and several esters, and glycerides that are modified from the cheap substrates like palm oil. Those products are broadly used in the various industries such as- chemical, pharmacy and food industry.

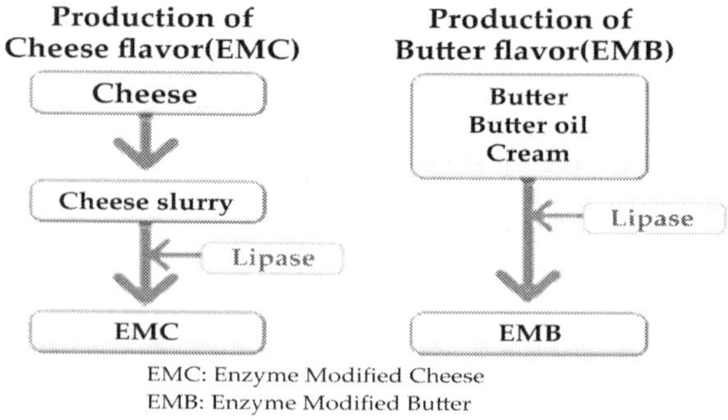

Figure 7.7: Production of cheese and butter flavourflavor by lipases

Source: https://www.amano-enzyme.co.jp/eng/productuse/oil.html

Proteases

To develop flavor in fermented milk products, it is essential that the proteolytic system of lactic acid bacteria grows in milk. The proteolytic system comprises of proteinases that primarily cleaves the protein present in milk into peptides; the peptidases cleaves the peptides into smaller peptides and amino acids, further for the cellular uptake of these small peptides and amino acids, the transport system is responsible. The proteolytic system of lactic acid bacteria is very complex, it is proficient for the conversion of milk casein into free amino acids and peptides that are essential for their growth. These include- extracellular proteinases, aminopeptidases, endopeptidases, tripeptidases, etc. all of which are serine protease. Other than the lactostreptococcal proteinases, existence of non-lactostreptococcal proteinases have also been reported. Since aminopeptidases are capable of releasing single amino acid residue from oligopeptides that are formed by extracellular proteinase activity, they are vital for developing flavourflavor in fermented milk products.

Amylases

Amylases are the derivatives of fungi and bacteria. There primary role is in food and beverage baking, brewing, and in sugar industries. Amylase is used for hydrolysinghydrolyzing amilum into a water soluble and low molecular weight product, named glucose. Amylases are widely used in drink industries for high-fructose syrup production, amylases are also used in textile industry as well. Amylases are made from diverse microorganisms such as- bacillus, clostrium, pseudomonas, etc. Bacteria's that are responsible for producing amylases on industrial scale includes- bacillus stearothermophillus and bacillus licheniformis. However, it is desirable to choose Bacillus stearothermophillus as this can produce thermos stable enzymes, and hence, reduces the cost of production at large scale.

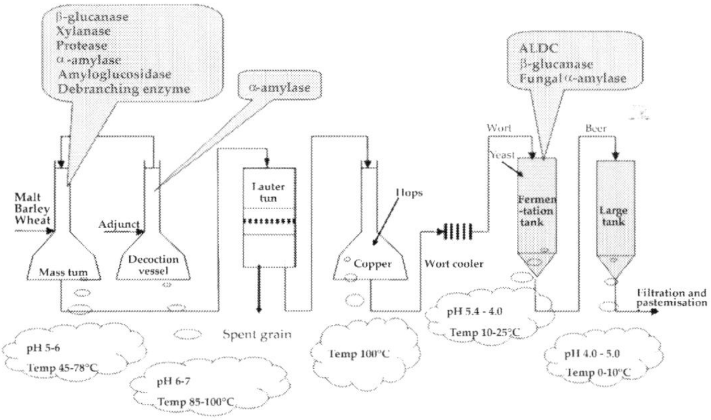

Figure 7.8: Amylases and other enzymes used in brewing process

Source: http://www.biokemi.org/biozoom/issues/522/articles/2368

It is important to note here that alpha-amylases have substantial effects on the baked goods. If the amount of amylase is lesser, this in turn results in lower dextrin production and poor gas production. This gives poor quality breads with poor crust color and reduced size. Hence, to overcome such shortage or deficiencies, it becomes obligatory to add sugar or alpha-amylase. Since alpha-amylases are the enzymes so they are advantageous over sugar to be used. Another enzyme named, cereal beta-amylase play a very significant role in releasing fermentable sugars from cereal grain starch to fuel. It is inspected that their amount is generally large due to their implication in economically important industry.

Glucose oxidase

Glucose oxidase enzyme is obtained from the fungi like such as Aspergillus Niger and Penicillium, this enzyme catalysescatalyzes the oxidation of beta-glucose to glucono-1,5-lactone by using molecular oxygen and releasing hydrogen peroxide. It is a highly specific enzyme and is used in removing oxygen or glucose from food products so as to expand their storage ability. Hydrogen peroxide is an effective bacteriosidebactericide that is removed after use, by treating with catalase which converts it to water and molecular oxygen. The two enzyme activities are not separated for most of the large-scale applications. To avoid the net hydrogen peroxide production, both glucose oxidase and catalase are used together. The principal application of glucose oxidase/catalase is in baking industry where it is used for removal of egg while before they dry out. To guarantee that sufficient molecular oxygen is available to oxidiseoxidize glucose, mixture of enzymes is used with extra hydrogen peroxide. The other uses of glucose oxidase are in removal of oxygen from the canned drinks and limiting the non-enzymic browning in mayonnaise and wine.

Figure 7.9: Baked foods in which enzyme glucose oxidase is used

Source: https://www.indiamart.com/proddetail/glucose-oxidase-10043460391.html

α-Acetolactate decarboxylase

Alpha-acetolactate decarboxylase isolated from the stain 1033 of Enterobacter aerogenes, which was then applied for beer maturation. By adding this enzyme to the beer, it results in removal of alpha-acetolactate and alpha-aceto-alpha-hydroxybutyrate within 24 hours after its addition, without affecting the natural taste and other properties of the beer. When the organoleptic properties of beer are compared, it is seen that the beer that is matured in alpha-acetolactate decarboxylase presence have satisfactorily equal quality when compared with the beer prepared conventionally. Studies shows that other than the removal of di-acetyl-2,3-pentanedione and the other ancestors of these compounds, there are hardly any events that are related to development of flavourflavor that occur during beer maturation. Subsequently, it is noted, that application of enzymes allows flavourflavor maturation to be carried out within 24 hours.

Cellulase

Cellulase enzymes are produced by bacteria, fungi, and protozoans, that aids in catalysingcatalyzing the cellulose hydrolysis. Cellulases are also produced by various other plants and animals too. The structure and mechanism of different cellulases also differs. Cellulose are breaking down into beta-glucose with the help of the enzyme, cellulase, such cellulase is produced by symbiotic bacteria in ruminant chamber of herbivores. Since many animals including humans don't produce cellulase enzyme in their body, therefore, they cannot use most of the energy that the plant materials contain. Another enzyme named hemicellulase is responsible for hydrolysinghydrolyzing hemicellulose, that are generally classified under cellulase, another enzyme that cleaves lignin is also classified as cellulase. Cellulase is widely used in commercial processing in coffee, in textile industry and in laundry detergents. Other uses of cellulase is in paper and pulp industry for many purposes, and also in pharmaceutical applications.

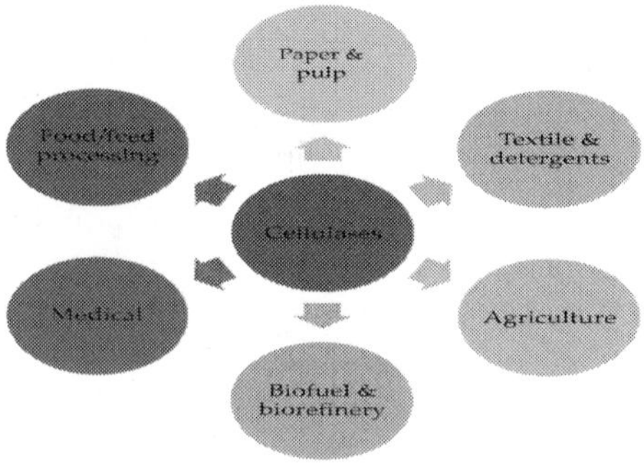

Figure 7.10: Cellulases enzymes uses in diverse fields

Source: http://www.sciencedirect.com/science/article/pii/S1043452616300353

Pectinase

Another important enzyme in the biological sector are the pectinases, and their use is increasing in the market continuously. Pectinases are the leading enzymes in the industry, they are the eco-friendly device of nature and are used in various industries such as food industry, wine industry, paper industry, in fruit and vegetable processing, animal feed processing, vegetable oil extraction and plant fibre sourcing. Since enzymatic catalysis is more specific, less aggressive and saves more energy, they are preferred more over chemical methods.

MAJOR APPLICATIONS OF ENZYMES IN FEED INDUSTRY

As discussed above, enzymes are the biological catalysts, they boost the nutrient availability of feed stuffs, lessen the feed cost and reduce the waste output in the environment. Although invention of enzymes in the animal feed application is a newly developed phenomenon yet it had gained a lot of importance in this field, just like in other fields such as detergents, textiles, brewing, baking, etc. The enzymes that are developed needs to be appropriate for the application they are designed for. For the production of industrial enzymes from the microorganisms, fermentation and extraction process is done. The enzymes that are obtained from the genetically modified microbes, helps in production of enzymes in large quantities which makes them economically viable. The feed enzymes should necessarily be strong enough to stand at different temperature and pH change. The feed enzymes need to have the capability to stand at high temperature to resist pelletization and to stay for longer shelf-life. To make the feed enzymes more heat stable various modifications have been done in them over the past years, for example- feed enzymes have been evolved from liquid to powder form, and from powder to granular form. To give full protection to the feed enzymes, granular technology has evolved, in this technology enzyme molecules are coated with an inert material, such as- cellulose and wax. Granular technology gives longer shelf life to enzymes as well as it is appropriate for pelletized feed.

The used of feed enzymes in European countries and Australia is from the decade, however, India is newly exposed to this. The reason behind this was that earlier combinations of enzymes in the west were primarily made for barley and wheat, and they were found to be inappropriate for Indian diets. The reason behind this delay was that there was no developed technology for the production of appropriate feed enzymes in the feed industry. Another reason was that since soya and corn were thought to be the ideal ingredients, there was no as such requirement was felt to use external enzymes. However, both soya and corn were very rich and high price sources, so to meet the economic needs other alternative feed sources are also being looking for. To overcome the price crisis of soya and corn, meat meal sources were used, but this was not assumed to be the permanent source, as introduction of animal protein may lead to complications of contamination in long run. Therefore, Indian farmers discovered some new feed sources as a substitute for soya and corn, these feed sources are- safflower, sunflower, rapeseed, etc. (these are for the protein sources); sorghum, millets and rice bran etc. (these are for the energy sources). To use these ingredients, they need to be supplemented only by substrate specific enzymes. The cell wall arrangement of these high fibre feed components shows that they cover huge sum of arabino-xylans, pectic polysaccharides and some cellulose. For instance, essential amino acids in sunflower meal are very high but due to their high fibre content they are restricted to be used. Use of definite enzymes such as- xylanase for xylan; pectinase for pectin; and cellulase for polysaccharides, could help in breaking down the fibrefiber releasing energy and improving the protein digestibility by accessing the protein well when the fibrefiber is broken down. By this method, the cost of feed lowers down, also reduction of prteinprotein level in the feed is possible. The scope of enzyme use in the coming era is gradually increasing due to their high performance and the availability of various enzyme products.

In India, one of the best enzymes named phytase has been caught. This enzyme is known since decades still feed industry couldn't use it due to its high production costs. As the result of biotechnology, it was in 1990s that the microbial phytase became commercially available. This enzyme is responsible for breaking down the indigestible phytic acid in oilseeds as well as in cereals, and helps in releasing digestible phosphorous. With the use of phytase the use of high priced supplemental inorganic phosphorous such as di-calcium phosphate has been reduced. Addition of inorganic phosphorous also leads to unnecessary excretion of phosphorous in manure, this poses an environmental concern, particularly in areas with rigorous animal production. Certain amino acids, proteins and minerals such as calcium, magnesium, zinc, and potassium are released by phytase, that are complexed with the phytate molecule (antinutritive factor, giving better performance in animals when hydrolysedhydrolyzed). Most of the farmers today use phytase in the feed to lessen the cost. An interesting observation was seen in the coastal region of Andhra Pradesh, where both poultry and freshwater aqua culture farming exists in the same area. As phytase are used in poultry to decrease the phosphorous content in the poultry litter, which ultimately results in lessening of phosphorous levels in ground water, the aquaculture farmers are also benefiting in regulating the algal bloom. This example shows the mutual benefits to both poultry and to aquaculture farming.

Two other enzymes, named protease and xylanase have also been introduced in the aqua feed when the drift of substituting fishmeal with soya was increasing. Since the demand of quality fishmeal with low contamination was very high, therefore it was not readily available. Aqua feed manufactures are seeing for opportunities for the use of phytase to release non-available phosphorous from deoiled rice bran, wheat and soya. The use of enzymes in cattle feed is almost nil because the cost of cattle feed is very low and the cost of enzymes are very high. Concentrated feed pellets can be made by combining xylanase and cellulase with bagasse, straw as well as with other agricultural wastes.

Currently, many enzyme formulations are accessible with diverse activities in the market. Though the enzyme formulations made are standardized, their mode of standardization varies from producer to producer, as it is almost impossible to compare dissimilar feed enzyme products. The usage of the enzyme is judged on the basis of their quality parameters. This quality parameter embraces the type of enzyme that is present, activity of the enzyme (unit such as- IU, PU, FTU, FYT), their production method, formation of product as well as its free-flowing nature for mixability. Since each unit has its own set of explanation it is very hard to alter one unit to the other unit. The characteristics and composition of the enzymes are dissimilar, as the enzymes that are used in the feed industry are produced by various diverse microorganisms. It is by the good quality control techniques and experience of the manufacturer that assures the reliable results. Only in-vivo technique can be used to compare the enzyme products, but since the pH values of each microorganisms is different, their optimum temperature of operation is different, their affinity to bind with feed is also different, therefore, it is not likely to progress in-vitro technique which can predict the in-vivo performance. In India, most of the enzyme products are directly imported at present or specific enzymes are imported and framed. As the enzymes are imported from outside this makes them very expensive to use. Only few companies produce enzymes in India, as the technology used and the seed organism used are not up to the mark. India needs to understand the various aspects of enzymes, so that India can also participate actively in producing enzymes itself in the near future, which will make them economically viable to use.

Although the function of all the enzymes is almost similar when they are supplemented in animal feed, their function is to enhance the raw-material digestion efficiency, nutritionists makes use of enzyme at different approaches. The characteristics of each enzyme differs slightly from one another and this also affects their dietary inclusion rate- their requirement for protective coating, their work in the animal's gut, as well as their overall efficacy. Each enzyme that functions within the animals have certain specific roles to perform, each of them acts on the substrate to commence the chemical reaction within the body, but only few are suitable to bind with a particular substrate- this is similar to how the key fits into a particular lock only, which in turn opens the lock. If wrong key is inserted into the lock, lock does not get opens. Same relation is between the enzyme and substrate, if correct enzyme comes to bind with the substrate, then product is formed. If wrong enzyme binds to the substrate, product is not formed. Some of the benefits of enzymes are listed below-

- They help in reducing the diet costs when ingredient price matrix permits.
- They enhance the feed efficiency as the effect of nutrient release, which in turn decreases the cost of meat per pound.
- They are also important in enteric health management. Diet rich in fibres changes the viscosity of the gut, and excess fermentation results in normality. Pathogenic bacteria also thrive with excess dietary nitrogen.

Enzyme	Function
Amylase	Utilization of starch and complex polysaccharides
Arabinase, Pectinase	Degradation of anti-nutritional factors, improves animals's energy level
Mannanase	Degradation of mannans, a non-starch polysaccharide (NSP)
Cellulase	Degradation of cellulose in plant feedstuffs
Keratinase	Degradation of feather keratin for production of feather meal
Lipase	Emulsification and utilization of oils and fats
Phytase	Increasing bioavailability of phosphorus, trace minerals and protein in feedstuffs
Protease	Utilization of animal and plant proteins
Tannase	Elimination of tannin
Xylanase	Hydrolysis of non-starch polysaccharides like pentosans
Galactosidase	Acts on- galactosides to reduce anti-nutrional factors
Glucosidase	Acts on-glucans (NSP) to unlock nutrients
Invertase	Converts sucrose to glucose and fructose and improves animal's energy level

Figure 7.11: Various enzymes and their roles in feed industry

Source: https://thefishsite.com/articles/importance-of-feed-enzymes-for-shrimp-and-fish-in-sustainable-aquaculture

Several types of enzymes are used in poultry feed, each with a certain exact function and role, which will result in abundant benefits for the poultry producer. Some of them are listed below-

Protease

Proteases are responsible for breaking down complex proteins into simpler and shorter ones, called as peptides, and amino acids, which are subsequently the building blocks of the protein. Proteases are proficient in taking proteins that binds starch within feed ingredients, this results in creating more of the energy that is found in starch accessible to the animal. Versazyme, the BRI's natural enzyme feed additive, is a broad-spectrum protease which destroys the substrates of many proteins. According to the 2001 trial it was illustrated that the chickens fed with 2 percent lower protein diet complemented with versazyme grew as like the birds that were fed with their normal diet holding more protein. There are almost more than thousand enzymes that are identified today, but only some of them are commercially available, and even out of those hundred only a small fraction of them are produced on an industrial scale.

Carbohydrase

Carbohydrases are those enzymes that breaks down complex fibres and starch into simpler sugars. Carbohydrases helps in expanding the digestibility of carbohydrates in the feed, which in turn increases the sum of animal energy needed to the muscle growth and development. Some of the categories of carbohydrases includes- amylase, that breaks down starch into glucose, fructose, maltose, and further simpler sugars; and xylanase, that digests the complex carbohydrates that are found in the dietary fibres. Though chickens naturally produce those enzymes that help in carbohydrate digestion, but they don't produce all the enzymes that are required for breaking down the fibre completely.

Amylase: Amylases are responsible for breaking down the starch molecule into simpler sugars, thus in turn providing energy to animals.

Xylanase and Glucanase: Both xylanase and glucanase are responsible for breaking down the non-starch polysaccharides, thus releasing energy from the fibrous portion grains and its by-products.

Phytase

Phytase enzyme is responsible for breaking down the phytate molecule, which results in formation of phosphorous that is found in plants but is naturally unavailable for animals, to release the available phosphorous for the growth of animal. Phosphorous being the most indispensable nutrient required for the maintenance, growth, and in cell and tissue repairing. Corns are the foremost elements of the poultry diet, with substantial amount of phytate in them. The main aim of adding phytase is to expand the rate of absorption of phosphorous in the poultry feed. The amount of phosphorous that the animal absorbs also has an environmental advantage, once it is excreted, there is a leakage of undigested phosphorous in the ground which ultimately makes its way to lakes, rivers, and oceans.

An interesting observation was seen in the coastal region of Andhra Pradesh, where both poultry and freshwater aqua culture farming exists in the same area. As phytase are used in poultry to decrease the phosphorous content in the poultry litter, which ultimately results in lessening of phosphorous levels in ground water, the aquaculture farmers are also benefiting in regulating the algal bloom. This example shows the mutual benefits to both poultry and to aquaculture farming.

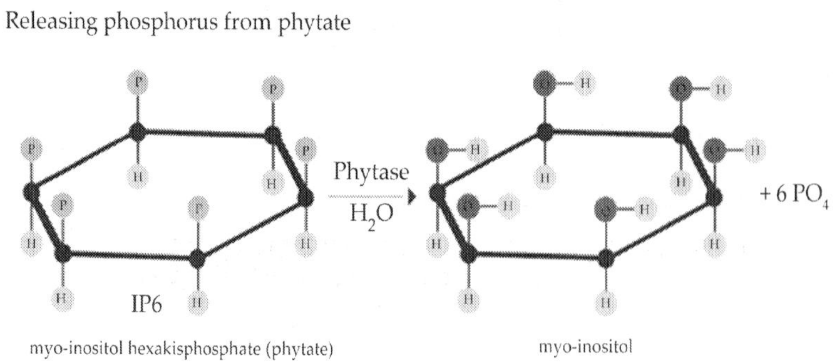

Figure 7.12: Releasing phosphorous from phytate with the help of enzyme phytase

Source: http://www.hubbardfeeds.com/blog/phytase-superdosing-beyond-ca-and-p-release

Enviropig

Enviropig is the genetically modified line of Yorkshire pigs, these pigs are capable of digesting the phosphorous in plant more proficiently than the normal Yorkshire pigs. The main aim behind choosing these pigs is that they produce phytase enzyme in the salivary glands, which is then secreted in the saliva. When these pigs consume the cereal grains, the phytase enzyme present in the salivary glands of the Yorkshire pigs' mixes with the feed as they chew. As soon as the food is swallowed, it activates the phytase enzyme in the acidic environment of the stomach, as the phytase enzyme is activated it degrades the indigestible phytate that is present in the feed which accounts for almost 50 to 75 percent of the grain phosphorous.

Since these Enviropig can digest the cereal grain phosphorous, the need to supplement diet with mineral phosphate or commercial phytase is negligible, also there is little phosphorous in the manure. This phosphorous depleted manure has an advantage to the environment, example- when this manure is spread on areas with high swine production, the probability of phosphorous to leach into freshwater ponds, rivers and streams is negligible. As phosphorous is mainly responsible to enable the algal growth, which kills the fishes because of the anoxic conditions created by excess phosphorous in the environment, the Enviropig are beneficial to overcome this condition as they produce very low phosphorous manure and hence result in reduced impact on environment. Due to these reasons, Enviropig are the biotechnology's one of the great inventions. The two principal characteristics of Enviropig are- they reduce the feed cost as well as the probability of water pollution.

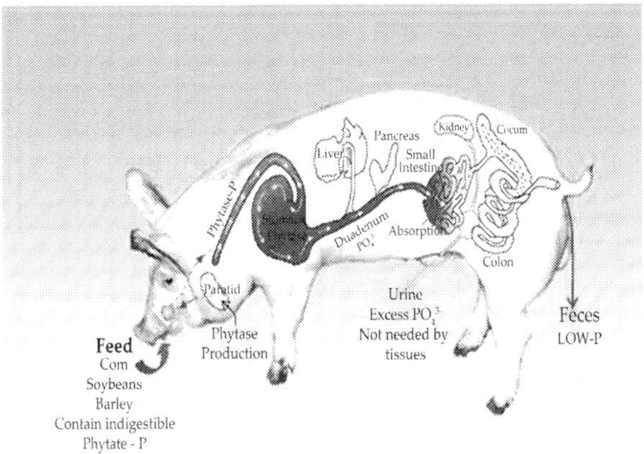

Figure 7.13: Phytase produced in the salivary glands and secreted in the saliva increases the digestion of phosphorus contained in feed grains.

Source: http://www.uoguelph.ca/enviropig/

REFERENCES

1. Canada.ca. (2014). Enzymes used in Food Processing - Canada.ca. [online] Available at: https://www.canada.ca/en/health-canada/services/food-nutrition/food-safety/food-additives/enzymes-used-processing.html [Accessed 29 Jul. 2017].
2. Houston, D. (2010). Enzymes and their applications in food » Houston Enzymes. [online] Houston-enzymes.com. Available at: https://www.houston-enzymes.com/learn/articles/enzymes-and-their-applications-in-food.php [Accessed 29 Jul. 2017].
3. Ltd., P. (2015). Enzymes and food additives. Enzyme and enzyme preparations for food, textile and leather industry. Bread improvers and food additives for meat processing, confectionary, production of mayonnaise and emulsified foods, tomato sauces and ketchup. [Online] Analytica-bg.com. Available at: http://www.analytica-bg.com/en/enzymes-food-additives/ [Accessed 29 Jul. 2017].
4. Porta, R., Pandey, A. and Rosell, C. (2010). Enzymes as Additives or Processing Aids in Food Biotechnology. Enzyme Research, 2010, pp.1-2.

CHAPTER 8

THICKENERS, STABILISERS, GELLING AGENTS, EMULSIFIERS AND FAT REPLACERS

"You don't have to cook fancy or complicated masterpieces- just good food from fresh ingredients."

-Julia child

It is important to keep the texture, consistency, and quality of food stuffs, such as ice-creams, mayonnaise, salad dressings, and other foods. For this, the manufactures use various kinds of thickening, gelling, stabilising, and emulsifying agents.

This chapter discusses about these agents and their uses in different food varieties. Apart from this, the chapter also deals with fat-replacers that are used as a substitute to conventional fats.

OVERVIEW

It is important to check the quality, texture, and consistency of food before their commercialization in the market. Thus, to maintain the consistency and texture as well as to avert the separation of ingredients in food stuffs such as butter, salad dressings, mayonnaise, milk, ice-cream and other dairy products, emulsifiers, thickeners, stabilisers stabilizers and gelling agents are used. Many reduced and low-fat versions of these have been developed with the help of these food additives. Furthermore, they are used in breads and other baked foods to support absorption of fact into the dough, resulting in improved volume, structure and a longer shelf-life. The texture consistency of different food additives is different when they are used as an ingredient. Therefore, to maintain consistency of required texture, precise emulsifiers, stabilizers, thickeners, and gelling agents are used. Some food uses gums as stabilisers stabilizers while others are dependent on phosphates or starch. For example, foods in which gums and starch are utilized utilized as stabilisers include- frozen desserts, salad dressings, cereal foods, snack foods, jams, jellies, pudding mixes and beverages. Different food additives also differ according to their food processing techniques. For instance, if one needs to serve food hot, stabilizer like pectin is used, on the other hand, it one needs to serve it cold, gelling agent like alginate is used. Since gelatine is a derivative of collagen, which are obtained from the by-products of animals, the vegetarian foodstuffs use agar-agar, guar, carrageenan and xantham, as an alternate for gelatine. Guar and xantham are often used in gluten-free foods as a substitute for wheat and related grains. Emulsifiers, thickener, stabiliserss and gelling agents are also used in the confectionary industry in the making of candies. For instance, to obtain the accurate consistency for moulding the chocolate into chocolate bars as well as to extend its shelf-life by postponing its onset of bloom due to which the chocolate turns white at the surface when stored at very high temperature, emulsifiers are added to the chocolates.

Generally, the emulsifiers, gelling agents, stabilisers and thickeners are odourless and tasteless. Emulsifiers, stabilisers, thickeners, and gelling agents are also used in the pharmaceutical and cosmetic industry. In the pharmaceutical industry, they are used as nutritional supplements and in various medications. In the cosmetic industry, their use mainly lies in personal hygiene products and haircare. They are generally used in creams, ointments, balms, moisturizers and pastes.

Stabilisers, thickeners, and gelling agents are obtained from various natural raw materials which are further merged with the food products to give a proper structure, stability, flow and eating qualities that the consumer needs. These stabilisers, thickeners and gelling agents all comprise of traditional materials like starch, that is a thickener obtained from various land plants; gelatine, which is a by-product of animal that gives characteristic melt-in-the-mouth gels; cellulose, which is a rich polymer in land plants, etc. To produce emulsifier, food additive, and food thickeners, gums play a very vital role in all the above. Gum technology stabilizers are the precisely measured mixture of numerous food ingredients. Over the past two decades, the understanding on gums and stabilisers and their use has increased enormously. Many of them are used as thickeners, gelling agents and stabilisers that occur in food naturally.

Thickeners- Thickeners are used to make the food thicker. Thickening agents are used to alter the effects of sugar in low-sugar or low-fat food products. To restore the texture of the food products from which the fats and sugars are removed, thickeners are used in place of them. For example, thickeners added to the thick milk shakes.

Figure 8.1: Milk shakes thickened by using thickening agents

Source: https://smoothiegains.com/thick-and-tasty-high-protein-milkshake-recipe/

Gelling agents- To maintain the structure and shape of the food gelling agents are used. Many of the fruits like- damsons, gooseberries, currants, bitter oranges etc. are rich in acid and pectin, therefore they are made into jams easily. Whereas fruits like- raspberries, strawberries, and cherries requires some additional amount of pectin to form jam. To form water-based jellies, desserts and confectionary jelly it is mandatory to add gelling agents as without them, it is impossible to make or set these foodstuffs. The soya protein used in vegetarian food products, may probably disintegrate at high temperature if gelling agents are not added to it. The property of gelling agents is to hold the parts together.

Figure 8.2: Gelling agents used in jellies

Source: http://www.ingrizo.com/en/confectionary-chocolates/glucose-syrups-thickening-and-gelling-agents

Stabilizers - stabilising agents are used to uphold both physical and textural properties of food products. The property of stabilizers is to obstruct the reaction occurring among two or more chemicals or amongst separation of emulsions, suspensions and foams. For instance, it comprises of surfactants and emulsifiers. Surfactants are the compounds that slows down the surface tension or interfacial tension between two liquids. Out of all the functions of stabilizing agents, one of the most important one is to act as thickening agent to gel the food products to the obligatory constancy. To uphold the texture of water-based products and food, stabilizers, thickeners, and emulsifiers all work together. The role of emulsifying agents is to stabilize the emulsion, which helps in forming a homogeneous mixture or two immiscible liquids that would not mix together otherwise.

Figure 8.3: Stabilizers used in ice-cream

Source: https://www.tateandlyle.com/sites/default/files/styles/article_medium_397 x210/public/Ice%20Cream.jpg?itok=zCo4cgNh

Emulsifiers- Emulsifying agents have both hydrophilic and hydrophobic molecules that keeps them away from mixing, and thus helps in separating the two liquids from one another by reducing the interfacial tension between them.

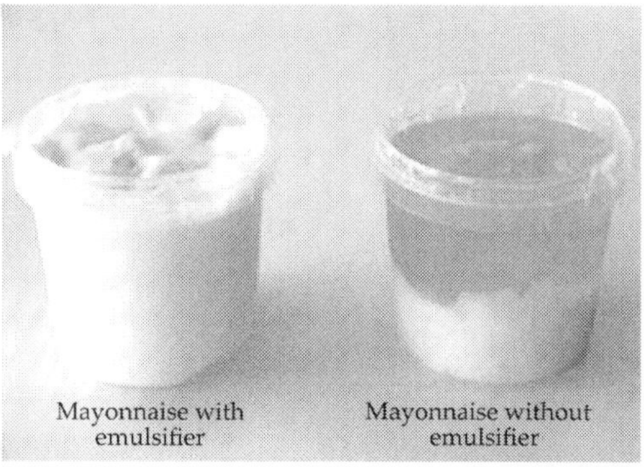

Figure 8.4: Mayonnaise with (left) and without (right) emulsifiers

Source: http://sarangminyak.blogspot.in/2013/11/emulsifier.html

Fat replacers- These are those substances that are used to substitute the fat that is consumed either partly or whole of it. Fat replacers are used to decrease the amount of fat and calorie intake in the food. Fat replacers are generally categorized under three categories, named carbohydrate-based fat replacer, protein-based fat replacer, and fat- based fat replacer.

The carbohydrate- based fat replacers absorbs water from the food and hence boosts its thickness anesstabilizes it. Some of the carbohydrate-based fat replacers include- guar gum, xanthum gum starches, oarefiber, dried plum paste, whearefiber, and modified foods. Foods products that contain carbohydrate-based fat replacers are frozen desserts, yogurts and cheese.

To make protein-based fat replacers, the protein molecules are broken down into smaller protein particles by the process of micro articulation. Milk protein or egg white proteins are required to make protein-based fat replacers. The protein-based fat replacer contains 4 grams in one calorie. Protein-based fat replacers are used in mayonnaise, butter, cheese, soups and salad dressings.

The fat-based fat replacers contain those molecules that are either absorbed partially or completely. For instance, in case of caprenin, benefit and olean; out of the three olean cannot be absorbed whereas benefat and caprenin are partially absorbed.

Figure 8.5: Fat free chips

Source: https://www.diet.com/g/fat-substitutesers

STABILIZERS IN FOOD

Stabilizers are the essential substances that are added to the food to uphold the physio-chemical state of the food produsersStabilizers are used in the food to give definite shape to the food and to maintain the texture and consistency of the food. Stabilising agents are added to the food stuffs in only certain specific amounts, neither less nor mosersStabilizers are used to provide uniformity to the food product as well as to hold the scatteringvourflavored compounds. The principal rolesersstabilizers is to persist a homogenous state even after merging of forming a blend. Sometimes food preservatives are confused with the term fsersstabilizers. However, the two are totally different from one anothsersstabilizers are used to hold the texture, consistency and other physical characteristics of the food whereas food preservatives are used to stop the spoilage of food product as well as to increase its shelf-life. The stability and thickness is increased by binding the fsersstabilizers with the larger molecules. The purposesersstabilizers is clear, they help in maintaining the consistency of the foodstuff and to maintain the ingredients such that they don't separate from one another.

Common isersstabilizers

Some of the most common isersstabilizers and the foods in which they are used are listed below-

Alginic acid

Alginic acid is one of the most popular type of food stabilizer which is a derivative of brown algae. The production of alginates mainly occurs by processing of brown sea-weed, *Macrocystis pyrefera.* The algin that is found in the kelp of the weed is present as a mixed salt with alginic acid. By processing, the water insoluble alginic acid is extracted and water-soluble alginates are formed. Sodium alginate, is the most widely used form of alginate. It has the capability to react with both divalent and trivalent metal ions which creates a chemically set irreversible gel which has many useful evidences in the food applications. Alginates are frequently used in thickening, suspending, stabilizing and emulsifying. Sodium alginates are used in the preparation of dry pudding mixes. Alginates can stop ice crystal formation while preserving the smooth texture and consistency. Alginates are also used in many other products like in beverages, whipped toppings, sauces, dressing, pie fillings and spreads. Nonetheless, alginates are not only restricted to foods, they are also used in various industrial applications like in paper coating, adhesives, textiles, air freshener gel and ceramics. Alginate has an outstanding function in terms of thickening agent, gelling agent, emulsifier, stabilizer, texture-improver, and to boost the food quality. Alginic isersstabilizers are used in jelly, lactic drinks, dressings, instant noodles, beer, ice-creams and dessert syrups. FAO/WHO has certified alginates as one of the safest food additives to be used in food products.

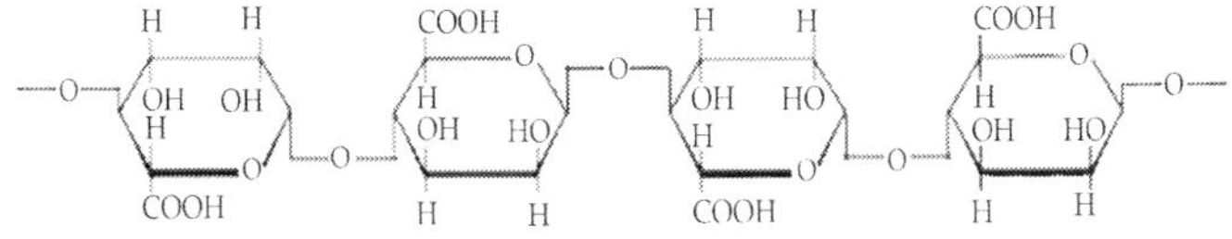

Alginic Acid

Figure 8.6: Chemical structure of alginic acid

Source: http://www.epharmacognosy.com/2012/03/algin.html

Agar-agar

Agar-agar is a natural product that is obtained from the marine red algae or sea-weed. Itlavourflavorless compound which when dissolved in water, heated or cooled becomes gelatinous. They form an opaque, brittle texture gel. Agar-agar is also recognized as Ceylon moss, china grass, Japanese isinglass. Agar-agar is a vegetarian alternative where gelatin cannot be used, as it is an animal by-product. Agar-agar that is available in the market are generally derived from *Gelidium amansii, Gracilaria species, or Sphaerococcus species*. Agar-agar is capable of prg colourcoldourlessodorless and natural gels without the need of any other colloid. Agar-agar forms gels at 34 to 42°C and it melts at 80 to 98°C, this quality of agar-agar is advantageous to increase the shelf-life of food preparations. Agar-agar has various uses in different industries like- it is used as a gelling agent in confectionary and food industry, for example as thickener in soups, ice-creams, yogurt and other desserts, as a clarifying agent in brewing industry as well as in the canned meat industry. The usage of agar-agar is increasing day to day as world is preferring and shifting more towards the products that are made from natural origin and don't include any animal sources. Agar-agar is one such example of the same which is used as a high-performing gelling agent since centuries.

Figure 8.7: Sago pudding made by using agar-agar

Source: http://foodcnr.com/sago-pudding-with-agar-agar/

Carrageenan

Carrageenan is a food additive extracted from re sea weed, *Chondrus crispus*, which is also called as irish moss. Carrageenan belong t sulphatesulfated polysaccharide family, and it is an indigestible polysaccharide. Although, carragennan don't provide any nutritional value to the food, they are still used in the food as thickener and emulsifier to enhance the ice-cream texture, improve consistency of yogurt, cottage cheese, soy milk and several other processed foods. They have been used in food preparations since decades, however, in the recent years it was found that carrageenan often leads to serious health concerns but there is no as such evidence regarding the same, hence their usage is continued in the food products. In the food industry, carrageenan is extensively used for their thickening, stabilising, and gelling properties. Particularly, carrageenan is used in non-dairy milks like coconut milk and almond milk. Usage of carrageenan is not only limited to the food industry, they have a wide usage in pharmaceutical industries too, in the pharmaceutical industry they are the active ingredients in the solutions and are commonly used to treat cough and intestinal problems. They are also used to treat peptic ulcers and also used as a bulk laxative. Carrageenan also play an important role in healthcare products like toothpaste and shampoos.

Figure 8.8: Carrageenan rich coconut milk, almond milk and other non-dairy products

Source: https://lifeholistically.com/carrageenan-is-it-ok-to-consume/

Gelatin

Gelatin is commonly made up of amino acids like glycine and proline. Gelatin is a derivative of collagen which is found in animal bones, fibrous tissues and various organs of animals, because of their origin from the animal source many people don't prefer to use gelatin in their diets and otherwise too. Gelatin is a clear and tasteless protein which is used in thickening of the food products. Befog gelatinegelatin to any of the food products, they need to be dissolved first in warm water, after dissolving they can be used in any liquid or semi-solid. GelatineGelatins are used in thickening the pudding, gummy candies, yogurt, ice-cream, marshmallows etc. Other than in food is gelatinegelatins are also used in personal care products, drug capsules, and in cosmetic industry. They are widely famous as the bae gelatinegelatin. GelatineGelatin needs an appropriate temperature at which melts or solidifies. G, gelatinegelatins let at the near body temperature. The solidifying properties and strf gelatinegelatin are ruinede gelatinegelatin is boiled. Some of fruits like pineapple, papaya, guava, also contains certain enzymes that obstructs the af gelatinegelatin to solidify.

Figure 8.9 : GelatineGelatin stabiliser used in candies

Source: http://positivelycharleston.com/gelatin/

Pectin

Pectin comprises of the methyl esters of polygalactouronic acid and salts likesuch as- calcium, ammonia, sodium and potassium and the molecular weight is upto 150,000 Daltons. Pectin belong to the group of complex polysaccharides that is found in the primary cell wall and middle lamella of plant cells. Pectin generally comprises a sugar residue, named,

D-galacturonic acid. Some of the leading polysaccharides in pectin includes- homogalacturonan, xylogalacturonan, rhamnogalacturonan I and II. Pectin are generally obtained by exacting the edible plant parts, chiefly the citrus peel and apple pomace, which are then undergone selective precipitation using salts or alcohol. Traditionally, pectin is used as gelling agents in broad varieties of fruit-based products which includes jams, jellies, marmalades, desserts, yogurts, and in bakery products. Pectin is also used as a stabiliser in some juice based drinks and also in some acidic protein beverages. In confectionary industry, they are used to give good gel structure to jellies. The range of pectin that is recommended to use is between 0.5 to 1.0 percent only. Usage of pectin is not just limited to food industry, they are also significant in various other industries like pharmaceutical and cosmetic industries.

Pectin are generally the primary component of all plants but their composition various in all plants which depends on their maturity, part of the plant they are obtained from, tissue as well as the growing condition of the plant. Higher amount of pectin can be obtained from citrus fruits and leguminous plants than from the cereals. Citrus fruits generally comprised of 60-70 percent pectin in their dietary fibre. Other than citrus fruits and leguminous plants, apples, oranges, apricots, and grapefruit also contains high amount of pectin. Some other pectin sources are cabbage, carrot, banana, and beets.

Figure 8.10: Pectin used in jam preparation

Source: http://indianapublicmedia.org/eartheats/pectin-naturally-sweetening-jam-jelly/

Lecithin

To confirm that our organs are working properly our liver produces a substance named lecithin, the deficiency of which lead to dysfunction of the liver. This lecithin is found both in plants and animals. Lecithin plays a vital role in the synthesis of various minerals and vitamins, mainly in vitamin A, D, E and K. Lecithin also plays a crucial role in production of cells as it supports the cells to accepts the nutrients they require. If enough of lecithin is not produced by the liver, person requires to take lecithin rich foods in their diets to avoid various liver related disorders such as fatty liver disease, inflammation, cirrhosis as well as hepatitis. In pregnant women, if lecithin is not present in sufficient amounts then the nutrients are not transmitted accurately from mother to the foetus. Some of lecithin-rich food sources are- cabbage, brussels sprouts, cauliflower, beans, leafy vegetables, eggs, milk, cheese, yogurt, and other dairy products. Lecithin found in soy is measured as one of the universal sources, this soy rich lecithin is found in soy beans, soy milk and soy cheese garnished on salads. Lecithin, as a food additive is used as fat emulsifier. Lecithin is used in baked goods as a replacement ingredient to lessen the egg and fat content. Lecithin is used in ice-creams, candy bars, salad dressings and margarines. The commercially used lecithin is made from the sunflower oil.

Figure 8.11: Butter spread containing soy lecithin

Source: http://www.huffingtonpost.com/amanda-greene/soy-lecithin-why-is-it-in_b_2891780.html

Figure 8.12: Canola oil having soy lecithin

Source: http://www.huffingtonpost.com/amanda-greene/soy-lecithin-why-is-it-in_b_2891780.html

THICKENERS AND GELLING AGENTS IN FOOD

Thickeners are used to upsurge the solutions viscosity without altering the other properties. Thickeners are also used to expand the suspension of other ingredients by which he stability of the product is increased. Though, thickening agents are commonly used in the food industry, still they are not only restricted to this industry, some other products that also uses thickening agents includes explosives, paints, inks etc. Cosmetic and personal hygiene products also contain thickening agents in them. Some thickeners form a gel, which gets dissolved in the liquid phase as a colloidal mixture forming a weak cohesive internal structure. Proteins and polysaccharides such as starch, pectin and vegetable gums are required for the making the food thickening agenula, is a flavourflavorless starch powder that is used as a thickener. Some of the starches that are used under this category are- corn starch, potato starch, arrowroot, katakuri starch, tapioca, and sago. Alginin, guar gum, xanthum gum, and locust bean gum are used as vegetable gum food thickening agents. The collagen, egg whites, gelatin and furcellaran are the protein based food thickeners. Sugar based food thickening agents includes agar and carrageenan. Some of the thickeners act directly on the protein that in present in the food product, one such example of this is, sodium pyrophosphate, this acts on casein in milk while instant pudding preparation. Depending on their physical and chemical conditions different thickeners are used for the different applications, as same thickener is not proper for all applications, because of the variation in taste, clarity and several other responses. For instance, arrowroot is a better choice for acidic foods than corn starch, as arrowroot loses its thickening strength in the acidic medium. When the pH level is below 4.5, guar gum is used to lessen the effect of aqueous solubility, and hence it also reduces the thickening ability. Tapioca or arrowroot is also preferred over corn starch when the food needs to be frozen, as it gives spongy texture to the frozen foods. Some thickeners are also used during the final stage preparation of some foods, since theseners have flavourflavor, they are not markedly stable and but due to their capability and convenience, they are extensively used. One of the primary ingredients that is used in the kitchen for thickening gravies, stews and gumbos is flours, before adding to the foods flours are cooked thoroughly so the food does not taste uncooked. Another thickening agents used in gravies, stews and sauces are roux, which is a combination of flour and butter. To thicken and support the consistency of soureal grains like such asoatmeal, farina, and couscous are used. In the Middle East and Eastern Europe, yogurt is used for thickening soups. Another thickening agent is tomato puree, this not only thickens the food b adds its flavourflavor to the foods. Professional cooks also use egg yolks to thicken the sauces, but use of this thickener requires a narrow temperature range, if egg yolks are overheated it directly ruins the sauce. Some of the nuts and glaces that are made up of meat or fish are also used in thickening the food products.

Thickeners like starch needs additional care in cooking, otherwise they will lose their thickening quality at high temperature or when they are cooked longer, also when give starchy taste when they are not cooked for desired duration. During cooking, the higher viscosity of foods burns them easily. It is suggested to use thickener in certain specific amount only so asntain the flavourflavor and texture of the food.

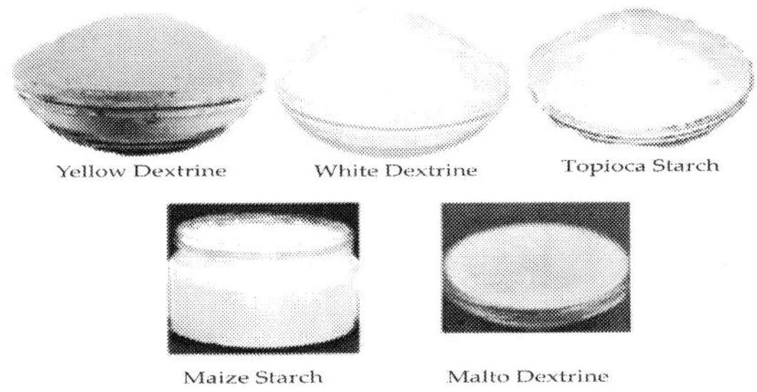

Figure 8.13: Diverse variety of thickeners used in foods

Source: http://www.sherskitchen.com/differences-among-all-the-various-types-of-food-thickeners/

To upsurge the thickness and to stabilise the food, gelling agents are used. They are used in various foods such as jellies, candies and desserts. By forming a gel, gelling agents provide appropriate texture to the food. Gelling agents are commonly trs and stabilisersstabilizers. Some of the typical gelling agents are pectin, natural gums, agar-arches and gelatinegelatin. Gelling agents are habitually based on proteins or polysaccharides.

EMULSIFIERS IN FOOD

Emulsifiers are those molecules which have both hydrophilic and hydrophobic end. The hydrophilic end in food is water friendly whereas the hydrophobic end is oil friendly. Emulsifiers that are useare also recognisedrecognized as emulgents. Milk, is the classical example of emulsion, in which the complex fat mixture is suspended in the aqueous solution. Both proteins and phospholipids are the examples of natural emulsifiers. Another most commonly used emulsifier is egg. Many emulsifiers such as magnesium stearate, sodium, potassium, and calcium salts of fatty acids acts like anti-caking agents. Sorbitan monoste, acts both likefier and stabiliserstabilizer in foods. The raw materials that are often used as emulsifiers includes- rapeseed oil, palm oil, sunflower oil, soy bean oil, etc. The combination of triglyceride with glycerol which ultimately results in monoglyceride is the most basic emulsifier that is produced. The nature of emulsifier obtained is governed by the kind of triglyceride used. For example, saturated triglycerides produce solid structures such as butter, and unsaturated triglycerides results in fluid products such as oil. In order to increase the emulsifying properties, monosaccharides are merged with someer substances, likesuch as- lactic acid and citric acid. Emulsifiers are also obligatory in maner industries like such as drug industry as well as in cosmetic industry.

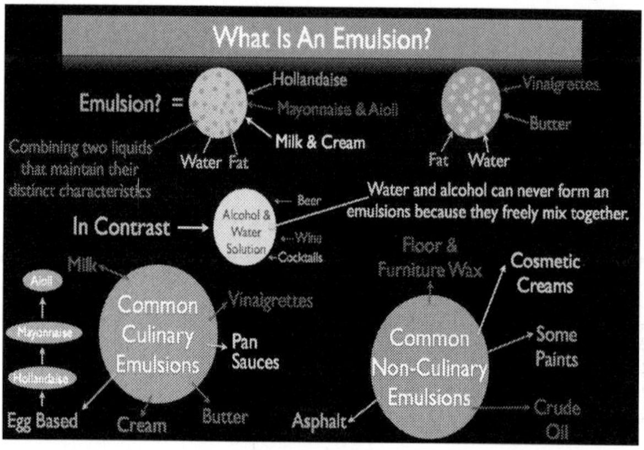

Figure 8.14: Understanding of emulsification

Source: https://www.slideshare.net/thyrnir/emulsification-pdf

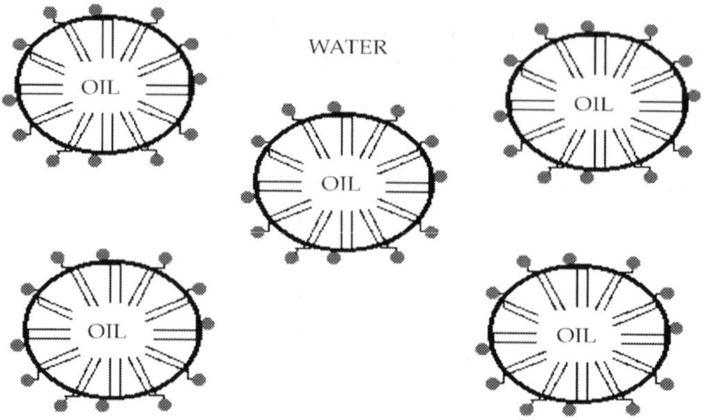

Figure 8.15: The water (hydrophilic end) and oil (hydrophobic end) in emulsifier

Source: http://www.huffingtonpost.com/amanda-greene/soy-lecithin-why-is-it-in_b_2891780.html

Emulsifiers are divided into four categories based on their hydrophobic groups, these are- anionics, non-ionics, cationic, and amphoteric food emulsifiers. Emulsifiers are added to the food products to give them look more alluring and attractive in appearance, as without emulsifiers both oil and water in food separates which makes the food look very unpleasant and non-tempting, not only this, emulsifiers also give freshness to thd products. The mouldmold that are formed on the foodstuffs if kept for too long is also prevented by natural food emulsifiers. Some of the food products in which emulsifiers are used includes- creams and sauces, some bakery and dairy products also contains emulsifiers.

Lecithin, mono-glycerides and di-glycerides of fatty acids, eaters of fatty acids as well as phosphate monoglycerides are some of the most common examples of emulsifiers. Some of the elementary food products in which natural food emulsifiers are used includes- biscuits, toffees breads, cakes, soft-drinks, extrudes snacks, frozen desserts, margarine, coffee whitener, and caramels. Mayonnaise, is also one such example of an emulsifier, here the emulsifier is phospholipid which exist in the egg-yolk. Almost 80 percent of the oil is dispersed in the aqueous phase in these successful emulsifiers. Another food product that cannot exist without emulsifier is ice-cream. Ice-cream is both foam and emulsion, the texture of ice-cream is formed by ice-crystals and the unfrozen water content. It's a myth that emulsifiers are not present in the creamy parts, some products where solid particles are also dispersed in airy foam, example bread and some baked products are also boosted by emulsifiers. Some of the natural and synthetic sources from where emulsifiers come from are illustrated below-

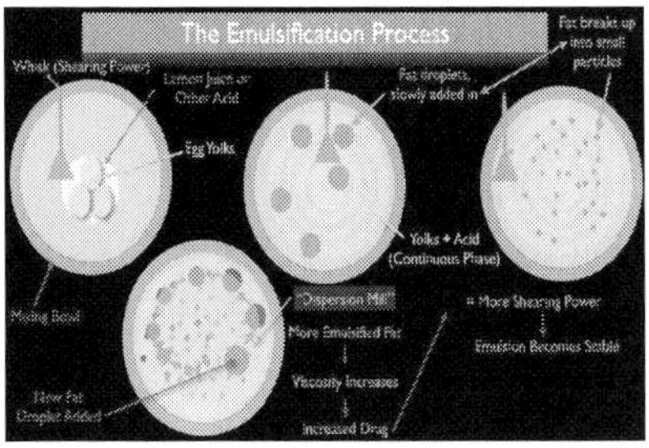

Figure 8.16: The complete emulsification process

Source: https://www.slideshare.net/thyrnir/emulsification-pdf

Esters of monoglycerides of fatty acid

The E-number of esters of monoglycerides of fatty acids is E472a-f. Natural fatty acids, glycend organic acids like such as acetic acid, citric acid, lactic acid and tartaric acid, together helps in the formation of esters of monoglycerides of fatty acids. Both vegetable and animal source are used in fatty acids, but vegetable source is normally preferred more. For example, collaboration between monoglyceride and ethylene oxide produces a resultant names ethoxylated monoglycerides. They are used in products like ice-cream, crisps and cakes.

Monoglycerides and diglycerides of fatty acids

The E-number of monoglycerides and diglycerides of fatty acids is E471 in the European Union, in United States, they match with the number 182.4505, and in Canada they are equivalent to M.4 and M.5. Natural fatty acids and glycerol helps in the formation of monoglycerides and diglyceride of fatty acids. They are commonly obtained from both plant and animal sources and are the semi-synthetic emulsifiers. Food products in which these emulsifiers are used includes- cakes, breads and margarines.

Fatty-acid derivatives

Some of the most common food emulsifiers that are derived from fatty acids are polyglycerol esters (PGE), stearoyl lactylates, propylene glycerol esters (PGMS), sucrose esters, sorbitan esters and polycarbonates. In food products like cakes and iceings, margarine and salad oils, Polyglycerol esters are used, PGMA is primarily used in whippable topping. In breads, stearoyl lactylates main application is in strengthening and conditioning the dough. For aeration in cakes and icings, both sorbitan and polysorbates are used. Foodstuffs sucha as bubble gum, soups, sauces, and liquid coffee uses sucrose esters.

Lecithin

The E-number of this emulsifier is E322 in the European Union and it is equivalent to L.2 in Canada. Since 1930s, lecithin is used as a food emulsifier in variety of foodstuffs. Lecithin is formed by combin of phospholipids like namely phosphatidylehanoamine and phosphatidyl choline. Phospholipids exact composition is dependent on its source. Soy beans and egg-yolks are the rich sources from where lecithin is extracted. Lecithis are generally used in bubble gums, baked goods, margarine, cakes, sauces, salad dressings, and chocolates.

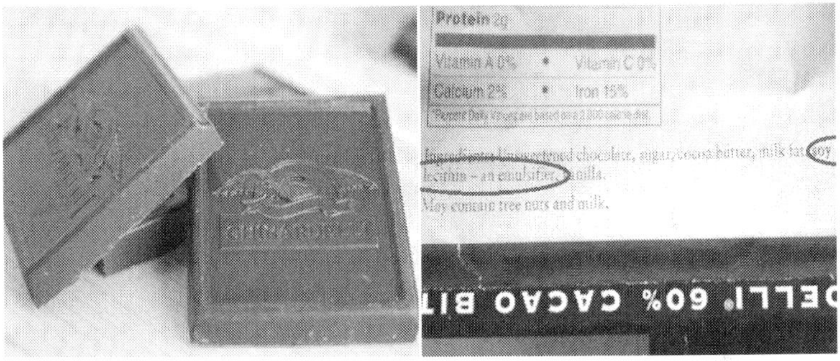

Figure 8.17: Soy lecithin used in chocolates

Source: http://www.huffingtonpost.com/amanda-greene/soy-lecithin-why-is-it-in_b_2891780.html

Lecithin is the most effective drug used medicinally for the treatment of high-cholesterol. Other than for treatment of high-cholesterol, leis also used for tratingtreating other healthted disorders too, like such as multiple sclerosis, alzheimer's disease and bipolar disorders. As lecithin is approved as asafe additive by FDA, maany of the food stuffs contains lecithin. The fatty compounds that are found in all the cells that exist in nature are lecithins. The hydrophobic wallell membranes are fromedformed by lecithin, also lecithin plays a significant role in brain functioning and in transporting some of the fat solussential nutrients like namely vitamins A, D, E and K, all over the body. Though, soy beans are the richest source of lecithin, some other sources i.e. corn, egg yolks, liver, wheat germ and peanuts also produces lecithin. Lecithin is composed of phospholipids, which are the three types of fat-soluble molecules. Lecithin is produced by the liver and is large intestine's most imporant component. Lecithins along with trigyceride and cholesterols, assists in facilitating cell communication and in keeping the cells from stickin ach other. Lecithis are is considered as the predominant building blocks of the cell membrane. Soy rich lecithin granules are produced by the lecithin rich soy beans, which is a natural compound which almost contains all the elements that are naturally found in the cell-membrane.

Lecithin can be found in variety of food products, as mentioned above. The principalrties of lecithin like- as surfactant, makes it an ideal emulsifier to be used in the food industry. Other than as emulsifier, lecithin also acts as wetting and gelling agent in some food stuffs. As lecithin is and emulsifying agent, therefore its predominant role is to save the ingredients from separating. For example, to preserve the foods that sticks to the pan, lecithin is added to the cooking sprays to avert this process. Many of the non-dairy products also holds lecithin, such as milk powder, sauces and gravies salad oils, margarine and shortening, and candy products. Lecithin usage is not just restricted to food industry, they play a vital role in therapeutic industry too, some of the main roles of lecithin in therapeutic industry are described below:

Effect of lecithin on liver

Lecithin is advantageous for treating alcoholic-induced cirrhosis. Cirrhosis is the condition in which the chronic liver cells swells up and causes bulid-up of scar tissue. The elevation of choline level in the liver which in turn rises the effective enzyme that helps in breaking down the collagen are the lecithin granules. Lecithin is also beneficial fro preventing fatty liver disease since lecithin has the capability to work as emulsifier and in breaking down fat.

Effect of lecn cardiovascular diseaesdisease

erve and muscle fucntionfunction can be improved by the use of soy lecithin granules. According to the Nutritional Supplement Education Centre, "without lecithin the membranes in our body would harden and our cells would die." Lecithin helps in mixing the fats with water and other body fluids, which breaks down the cholesterol and lipids, and helps in removal from the body effortlessly. This removal from the body, helps in preventing building- up in the arterial walls, serving the cardiovascular health and in prevention of atherosclerosis.

A 2009 joint clinical study by the Departments of Pharmacy at various universities in San Paulo, Brazil evaluated patients with elevated blood cholesterol. The results, published in the journal "Cholesterol," showed that daily administration of soy lecithin significantly reduced total cholesterol and stimulated fatty acid secretion with high-levels of cholesterol when compared with diets without lecithin.

Effect of lecithin on brain

As soon as the person consumes soy lecithin or it granules, before entering the cell-membrane it gets converted into nutrient choline. This choline in turn helps in the formation of acetylcholine, which is a neurotransmitter that helps in the proper functioning of the brain. Lecithin use is still indecisive in many of the related disorders like such as dementia, bipolar disorders etc. A study published by CNS drug review, used lecithin's derivative to investigate the aids of lecithin for analgesia and dementia therapy, by the studies they reach to the conclusion that lecithin is the most potential candidate for treating dementia related pathology.

Effects of lecithin on gall-bladder

A small organ where bile is stored and concentrated before its release in small intestine is gall-bladder. Gallstones that are formed in the gall-bladder helps in digestion. If the cholesterol level is very high it forms stones. The chief part of bile is lecithin, which aids in breaking-down the fat molecules by allowing them to combine with water, confirming their rapid elimination from the human body. The risk of gall stones elevates with age, diet, weight, and the hormonal changes in the body, as concluded by Healthier Life Website. According to the survey conducted, approximately 20 percent of the people who are aged over 65 develops gallstones, yet only 4 percent of them proves its symptoms. Lecithin which is a combination of choline, fatty-acids, inositol and phosphorous, performs many functions in the body. The concentration of lecithin is high in brain, liver, kidney, and bone marrow. Variable quantities of lecithin are found in all living animal and vegetable cells.

Lecithin granules can be ed in many forms, like- they can be taken directly in the mouth and chew until lecithin dissolves completely, other way to take lecithin is by adding a tiny-pellets of it to the mixer while preparing a smoothie, or by sprinkling them on soups and cereals. Another alternative is to cook the supplements in dishes or mixing the lecithin with cold or hot foods, as properties of lecithin are not affected by rise or fall in temperature. Some of the opposing reactions that occur by consuming high dose of lecithin are- liver inflammation, gastrointestinal illness, decreased appetite, nausea as well as extreme salivation. People who feel sick by consuming lecithin granules they should at once stop its usage, and try for some other alternative other than lecithin.

Emulsifying products and their usage

Some of the most common chemicals and their uses are discussed below in detail-

Enzymatically decomposed soybean phospholipid

Enzymatically decomposed soybean phospholipids are the improved soybean phospholipids which are the good natural food emulsifier. These are commonly used in biscuits, candies, ice-creams, margarine foods, and cakes.

Ammonium phosphatide

Ammonium phosphatides are the combinations of ammonium salts of phosphorylated glycerides which can either be made artificially or from glycerol blends as well as from partly hardened rapeseed oils. They are the lecithin substitutes and are frequently used as an emulsifier in chocolate industry.

Xylitan monistearate and tripolyglyceryl monostearate

Both the emulsifiers are used to enhance the taste and quality of canned drinks and to increase the shelf-life of preservation. They also increase the drink stability, by avoiding deposits, delamination, and formation of oil ring.

Polyoxyethylene (20) sorbitan mono-oleat, mono-stearate, sorbitan mono-palmitate, polyoxyethylene (20) sorbitan mono-laurate, sorbitan mono-oleate

These emulsifiers are used as a chemical intermediate in many industries. These emulsifiers are also usedther purposes too, like- as laming agent, solubilisingsolubilizing agent, washing agent, antistatic agent, degreasing agent, dispersing agent, etc.

Sorbitan tristearate

Sodium tristearate is used in many food and aerosol spraypersing agent, stabiliserstabilizer and as an emulsifying agent. It is a non-ionic surfactant, with a E-number E492, by the European Union.

Sorbitan monostearate

Sorbitan monostearate is an ester of sorbitan and stearic acid, which is also denoted as synthetic wax. The major use of sorbitan monostearate is as emulsifying agent.

Sorbitan mono-palmitate

Sodium mono-palmitate is principally used as an emulsifier as well as thickener in variety of food products.

Sorbitan mono-laurate

Sorbitan mono-laurate is used as an oil-in-water emulsifying agent. It is also used as stabilising agent, lubricant, dispersant, solvent as well as antistatic agent.

Calcium stearoyl lactylate, Sodium Stearoyl lactylate

Lactylate is the commercially available chemical. Calcium stearoyl lactylate is manufactured by bio-renewable feedstocks, and is non-toxic and biodegradable. Calcium stearoyl lactylate is considered as a very effective and safe food additive, hence used in broad variety of food products ranging from baked goods and desserts to packaging.

Sucrose esters

Sucrose fatty acid esters comprises of sucrose as hydrophilic group and fatty acid as lipophilic group, these are the non-ionic surfactants which are commonly called as sugar esters.

Octyl and decyl glecerate

Citric and fatty acid esters of glycerol (CITREM)

There uses includes emulsification of the fat products, dairy products, grease, meat and improving its stability, texture, and consistency.

Diacetyl tartaric acid esters of monoglyceride and diglyceride

To build a strong gluten network the strength of the dough is increased by diacetyl tartaric acid esters of monoglyceride and diglyceride. Their uses in the food industries are in giving chewy texture to food, crusting the breads, in ice-creams, coffee whiteners and in salad dressings.

Acetylated monoglyceride and diglyceride, sodium caseinate

Casein is one of the major components of the cheese used as a food additive and has variety of uses in other food products too, amino acids, carbohydrates, calcium and phosphorous are supplied by casein as a food source.

Propylene glycol esters of fatty acid

Propylene glycol esters of fatty acids are used to prevent the bread and confect aging, in pastry margarine, as well as in enhancing the processing performance. Food products in which propylene glycol esters of fatty acids are used includes- ice-creams etc., this helps in keeping the preservation of shape and to increase the dilatability.

Distilled monoglyceride

Distilled monoglyceride is a non-ionic emulsifier, and it works on broad pH range. It is used in many foods as well as in cosmetic industry. In cosmetic industry, it is included in body-cream, and hair products. In food industry, it is used in powdery bottom honey, powdery frost etc.

FAT-REPLACERS IN FOOD

Fat donates sensory and physiological benefit as a food component. The major contributiat is in proviflavourflavor, aroma odourodor, and taste to the food. Another important contribution of fats includes- it gives creaminess, palatability, appearance, texture, and lubricity to foods, also it gives a satisfactory feeling during the meals. Fats act as a precursthe development of flavourfr to stabilise the flavourflavor,carries lipophilic flavourflavoring compounds. Facts also acts as a precursor for prostaglandins, source of some essential fatty acids, fat soluble vitamins and as a carrier for lipophilic drug. Fats give approximately 9 kcal/g, which is highest compareother energy sources like such as proteins and carbohydrates 4 kcal/g. Person who consumes large amount of fat in their daily diet have higher chances to face obesity, coronary heart diseases, increased blood cholesterol rate, and some types of cancer.

Figure 8.18: Potato chips with varied fat content

Source: https://www.youtube.com/watch?v=2OrEPgiHLLE

According to USDA and USDHHS dietary guidelines in 1995, it was suggested to limit the fat intake to 30 percent of total energy intake daily, of which the percentage of saturated fat intake is 10 percent and two-thirds of are monsaturated and polysaturated fats. According to the survey conducted in America, it was specified that 56 percent of people have moved their interest from normal fat towards the food containing fat replacers, compounds with reduced amount of fat. Another survey directed by the Calorie Control Council in Atlanta, it was confirmed that people are more curious about low-fat, reduced fat or fat free food and beverage consumption rather than consuming the fat-rich foods. Even after the development of low-fat or reduced-fat foods in the market, there are high frequency of the population with increased weight or overweight, this is because people do not consume it in the amount as recommended and high fat consumption ultimately results in increase in weight and is the invitation to various other diseases in the body. To overcome such difficulties, fat-replacers are the best alternative to the health-conscious individuals, these fat-replacers helps in maintaining the basic food selection pattern and more easily adhere to a low-fat diet (CCC, 1996). Traditional techniques like usage of water or air to replace the fat, lean meats are used for frozen entrees, for frozen desserts instead of whole milk, skimmed milk are used and to prepare snacks baking is preferred over frying, all these methods aids in replacing the fats with fat-replacing entities. Also, by reformulation with lipid, protein and carbohydrate-based ingredients fats are replaced. Fat-replacers used in the foods have variety of functions and sensory properties as well as physiological effects.

Types of fat-replacers

Different authors use different definitions to describe fat-replacers. Those compounds which generally and chemically resembles fats, carbohydrates or proteins are known as fat-replacers. These fat-replacers are categorised into two groups, named as- fat-substitutes and fat-mimetics.

1575 Kcal
High Energy Density

1575 Kcal
Low Energy Density

Figure 8.19: Food products with high energy density (left) and low energy density (right)

Source: https://wedishnutrition.wordpress.com/tag/fat-replacers/

Fat substitutes

Those macromolecules that resembles to triglycerides both physically and mentally, are termed as fat substitutes. They generally replace the fat on gram-to-gram or one-to-one basis. Fat-based fat replacers and lipid-based fat replacers are often referred to as fat substitutes. Fat-substitutes are generally stable at frying and cooking temperatures. Some of the fat-substitutes are discussed below-

Sucrose fatty acid polyesters (SFP)

The combination of sucrose esters that are designed by chemical interesterification or transesterification of sucrose with six to eight fatty acids are known as the sucrose fatty acid polyesters. The exchange of an acyl radical between two esters is called as interesterification, whereas the exchange of acyl radical between esters and acids is called as transesterification. Olestra, is a sucrose ester that is formed from saturated and unsaturated fatty acids of chains with twelve of higher carbon atoms, olestra is attained from vegetable oils and edible fats. The fatty acid composition and degree of esterification are the specifications of olestra. Olestra's predominant role is to direct the functionality, potential applications, and physical properties. In 1996, FDA permitted olestra to be used in potato chips, cheese puffs and crackers as a 100 percent replacer for conventional fats. It isicult to absorb or metabolisemetabolize olestra, because of its large-size. Olestra has very high probability to caastrointestinal effects like such as abdominal cramping, softening, or loosening of stool etc., as they pass the tract without being digested. Due to the harmful effects of olestra on health, food products in which olestra are present are labelled as "This food contains olestra", by FDA. The absorption of other macronutrients i.e. carbohydrates, proteins or water-soluble vitamins and minerals are not suggestively affected by olestra.

Sucrose fatty acid esters (SFE)

These are prepared in the same manner as sucrose esters are made. They are made are the mono, di and tri- esters of sucrose with fatty acids. Sucrose fatty acid esters have caloric value as imply absorbed and hydrolysedhydrolyzed by the digestive lipases. Sucrose fatty acid esters are outstanding surfactants and emulsifying agents, because of the balancing of the five to seven free hydroxyl groups with one to three fatty acid esters. In United States, sucrose fatty acid esters are permitted to be used in variety of food prn emulsifying and stabilisingstabilizing agent. Other than their uses in the food industry, sucrose fatty acid esters are also used in thinning agents, anticaking agents, lubricants and in antimicrobials.

Carbohydrate fatty acid esters and polyol fatty acid esters

Both of these compounds are used as fat replacers in variety of food products. When one or more fatty acid esters react with the polyol which contains four hydroxy groups in the presence of a basic catalyst, they form polyol fatty acid esters. Sorbitol, raffinose, trehalose and stachyose polyesters are the few examples of sucrose fatty acid esters. The combination of tri-, tera-, and penta-esters of sorbitol and their anhydrides with fatty acids, forms a sorbitol polyester named as sorbestrin. Sorbestrin has a calorie value of 1.5 kcal/g. Due to the higher stability of sorbestrin than can withstand the frying temperature well. There is no commercial existence of sorbestrin yet, it is used in baked goods, salad dressings, and frying.

Alkyl glycoside polyesters

Alkyl glycoside polyesters are used as a substitute for conventional fat formulations. These polyol fatty acids are still under process and research is going on in lab.

Structured lipids

Triglycerides which contain either short chain fatty acids (SCFA), medium chain fatty acids (MCFA), and long chain fatty acids (LCFA) are known as structured lipids. By chemical or enzymatic synthesis or by random esterification these structured lipids are prepared. To reduce the amount of fat that is available for metabolism structured lipids are used. These structured lipids have very precise purpose.

Medium chain triglycerides

The saturated fatty acids with chain length of C8:0 (caprylic) to C10:0 (capric) with traces of C6:0 and C12:0 fatty acids are known as the medium chain triglycerides. Vegetable oils, like coconut oil and palm kernel oil, by rapid hydrolysis and fractionation of the fatty acid, and again re-esterification with glycerol to form triglycerides is used for the manufacturing of medium chain triglycerides. Both the conventional and functional properties of medium chain triglycerides are different from one another, the chemical structure results in functional properties that is dissimilar to conventional oil and fat. At higher temperature, medium chain triglycerides comprising saturated fatty acids do not undergo oxidation rapidly, and they are also stable at lower temperatures too. The solubility of medium chain triglycerides is higher than long chain triglycerides. The calorie value they provide is around 8.3 kcal/g. some of the basic uses of medium chain triglycerides inroducts o carry flavourflavor, colourcolor, and vitamins. Other than in food industry, they have many uses in confectionery industry too, like- they are used to give glossy feeling and preventing stickiness in confectionery products etc. Since 1950s, persons who suffer from absorption, digestion, or transport disorders, generally use medium chain triglycerides in their enteral and parenteral diets.

Caprocaprylobehenic triacylglyceride (caprenin)

Glycerol by esterification with caprylic (C8:0), capric (C10:0), and behenic (C22:0) fatty acids helps in the manufacturing of caprenin. The approximate calorie value that caprenin provides is 5 kcal/g, as the caprylic and capric fatty acids are readily metabolised with longer fatty acids and behenic acid is partially absorbed. Caprenin's properties resembles with cocoa butter. Some of the common uses of caprenin are in soft candies and also in confectionery coatings. Mixture of both caprenin and polydextrose is used commercially in chocolate bars, a low-calorie and low-fat compound.

Salatrim

The mixture which contains both short chain fatty acid and long chain fatty acid randomly attached to a glycerol molecule is known as salatrim. The calorie value of salatrim is approximately 55 percent or 5/9 the value of conventional fat, as the calorie value of short chain is lower than the calorie value of long chain fatty acid. Salatrim is developed by Nabisco Foods Group and licensed by Cultor Food Science, which gave them the brand name Benefat™ for manufacturing and marketing purpose. In 1994, FDA permitted the use of salatrim in food products. Salatrim works on varied range of melting points, they give hardness and appearance to food stuffs. Some of the basic food products in which salatrim is extensivelare- deposited chips, flavourflavored chips, caramel and toffees, peanut spread, savorygs, dips and sauces, fillingsfillings, and inclusions for confectionary and baked goods, and in dairy products such as cheese, sour cream, and frozen dairy desserts. Benefat 1, is the first commercially used salatrim product, which was prepared as a substitute of cocoa butter in many confectionery applications.

Dialkyl dihexadecylmalonate (DDM)

The fatty alcohol dicarboxylic acid ester of malonic and alkylmalonic acid which is synthesised by malonyl dihalide with fatty alcohol is known as dialkyl dihexadecylmalonate. To increase the molecular weight of dialkyl dihexadecylmalonate, a basic solvent named, alkyl halide is used. To replace oil in food formulations or in frying, Frito-Lay, Inc. has patented dialkyl dihexadecylmalonate for their use. Since, dialkyl dihexadecylmalonate cannot be absorbed or digested, it does not have calorie value. Talking about the commercial availability of dialkyl dihexadecylmalonate, it is still under process and not yet marketed.

Esterified propoxylated glycerols (EPGs)

Esterified propoxylated glycerols belongs to propylene oxide family. When glycerol reacts with propylene oxide, polyether polyol is formed which afterwards esterifies with fatty acids, and ultimately forms esterified propoxylated glycerols. The conventional triglycerides and esterified propoxylated glycerols are not similar in oxypropylene group positioning between the backbone of glycerol and fatty acids. In 1989, esterified propoxylated glycerols were patented by the company ARCO Chemical Co. and CPC International/Best Foods, to be used as a substitute of and oils in many foods like-such as frozen desserts, baked goods, salad dressings, for cooking, frying as well as for spreading. Esterified propoxylated glycerols have low calorie value as they are resistant to lipase. Again, the commercial availability of esterified propoxylated glycerols are not known.

Trialkoxytricrballylate (TATCA), trialkoxycitrate (TAC), and trialkoxyglyceryl ether (TGE)

The polycarboxylic acids with two to four carboxylic acid groups that are esterified with saturated or unsaturated alcohols with the carbon chain of eight to thirty atoms, that may be straight or branched are known as trialoxytricarballylate, trialkoxycitrate, and trialkoxyglycerol ether. These compounds are not susceptible to complete hydrolysis by lipases as the ester units are reversed by the corresponding ester that is present in triglycerides. The commercial availability of trialkoxytricarballylate, trialkoxycitrate, and trialkoxyglyceryl is not yet confirmed.

Fat mimetics

Those substances that resembles triglycerides in their physical properties but cannot replace the fat on gram-to-gram or one-to-one basis, are termed as fat mimetics. Protein-based fat replacers and carbohydrate-based fat replacers are often referred to as fat mimetics. Starch and cellulose are the most common food constituents, that are modified both physically and chemically to mimic the function of fat. Fat mimetics caloric value range is between 0 to 4 kcal/g. Extensive amount of water is absorbed by the fat mimetics. Fat mimetics are not stable when fried because extensive water binds with them, due to which at high-temperature they denature or caramelize. Some of the fat mimetics though are apt for baking and retorting. Fatcs have water-soluble flavourflavoer than lipid-soluble flavourflavors. Fat mimetics are broadly categorised under three main groups, as discussed below

Figure 8.20: Application of dairy fat replacer

Source: https://www.youtube.com/watch?v=zPueYU7OhVQ

Protein-based fat mimetics

There are certain substances that mimic the mouthfeel and texture of fat, they microparticulate to form microscopic coagulated round deformable particles, and are known as protein-based fat mimetics. Some of the most common sources of proteins from where the fat-replacers are derived includes- egg, milk, soy, whey, gelatin, and wheat gluten. To aingredient functionality, like-such as water binding and emulsification properties also, the protein-based fat mimetics are processed. Majorly protein-based fat mimetics are used in salad dressings, frozen desserts, dairy products, and margarines.

Simplesse, developed by NutraSweet Kelco Co., is one of the protein-based fat mimetics that is manufactured by microparticulation process of whey protein concentrate. In 1990, GRAS confirmed simplesse to be used in frozen dessert products. In 1994, again GRAS affirmed its usage in cheese spreads, yogurt, cream cheese, frozen desserts, and sour cream. Food producat do not require frying, like-such as dips, baked goods, frostings, mayonnaise, salad dressings, soups, and sauces, use simplesse as an additional product. Simplesse has a calorie value of 1 kcal/g. Like otheins simplesse provides flavourflavor to the food products, hence holds the biological value of the protein.

Carbohydrate-based fat mimetics

To replace fat partially or completely from food some food products use carbohydrate-based fat mimetics. The calorie value of the carbohydrates differs based on their digestibility, starch and dextrins which are digestible carbohydrates provide 4 kcal/g, and the non-digestible carbohydrates provide lesser calorie value than the digestible ones. Carbohydrates are also used for thickening or gelling the foodstuffs. Some of the carbohydrate ingredients like- gums, starches, cellulose, and pectin delivers the fat functions by binding with water. Like protein based fat mimetics, carbohydrates based fat mimetics also provide texture, mouthfeel, and opacity to the foSome of the fat replacers like-such as corn syrups, syrup solids, and high-fructose corn syrups are used in variety of foods as fat-free or reduced-fat cookies as a water activity (aw) controller. Carbohydrate based fat mimetics are used as a fat barrier in frying as well as baking. Some of the carbohydrate-based fat mimetics are-

Gums: Gums thickening agents, stabilisingstabilizing agent as well as gelling agent in variety of food product. They have a high molecular weight and are measured as negatively charged carbohydrate-based fat mimetic. There are many fat-replacing systems where gums are used with other gums or fat replacers or bulking agents, some of which are-xanthan, guar, carrageenan, locust bean gum, gum Arabic, and pectin. There are numerous uses of gums in salad dressings, dairy products, soups and sauces, ground beef, baked goods, desserts and ice-creams, icings and glazes, etc.

Starches: Fat-replacing system of varied starch sources, types, and functional sources are used to deliver sensory properties of oil. Some of the common starch sources are- waxy maize, corn and high amylase corn, potato, wheat, tapioca, waxy rice, and rice. Starch is modified by acid or enzymatic hydrolysis, dextrinization, oxidation, crosslinking, or by mono-substitution to reach the required functional and sensory properties. Starch is generally available in instant forms and they work better in fooith hig moisture content, likesuch as- salad dressings and sauces, margarine spreads, frostings and fillings, baked goods, etc. On the other , their performance is not uptoup to the mark for the foods with low moisture content, like- crackers or cookies.

Cellulose-based fat replacers are also used for replacing fats, they are of plant origin and are mainly obtained by chemical depolymerization, mechanical grinding, and wet mechanical disiwell as chemical derivitizationderivatization. As considered by GRAS, **microcrystalline cellulose** has no caloric value. In aqueous systems, microcrystalline cellulose mimics fats, contributes to the body, gives consistency, and mouthfeel, controls syneresis, stabilizes the emulsion and foams, adds viscosity, opacity and gloss the food products. Some of the important applications of microcrystalline cellulose in food industry are in- dairy products, frozen desserts, sauces, and in salad dressings. Another form of cellulose affirmed by GRAS is powdered cellulose, which is used in reduced fat sauces. Other uses of **powdered cellulose** are in- reducing the fat in fried butter coating as well as in fried cake donuts. Due to the unique propowdered cellulose to stabbilize air bubbles and minimiseminimize the shrinkage, they are used in elevating the volume of baked goods.

Maltodextrin- Maltodextrin is a safe, non-sweet, nutritive mixture of saccharide polymer of varying length as affirmed by GRAS. Corn or potato starch's partial hydrolysis produce maltodextrins. Some of the common foods from which maltodextrins are obtained are- rice, wheat, oat, etc. matodextrin's average molecular weight as well as the degree of hydrolysis is varied based on the dextrose equivalence (DE) of 20. Major application of maltodextrin is in contributing smooth mouthfeel in fat replacing systems, to build solids and viscosity, binding and controlling water, and browning activity.

Polydextrose- Glucose, sorbitol, and citric or phosphoric acid are randomly bonded into polymers to form polydextrose. Polydextrose provide calorie kcal/g, and is only metabolisedmetabolized partially. Polydextrose is officially used as formulating aid, bulking agent, texturizer, and humectant in variety of food categories, such as- baked goods and baking mixes, confections and frostings, salad dressings, chewing gum, frozen dairy desserts and mixes, pudding and fillings, gelatins, hard and soft candy, peanut spread, toppings and syrups, and sweet sauces.

Oatrim- Oatrim is prepared by the partial enzymatic hydrolysis of starch which is obtained from whole oat or corn flour. Oatrim is either used as a dry powder or as gel hydrated with three parts of water, oatrim comprises of 5 percent of beta-glucan. Oatrim is not appropriate for frying, but it can withstand at high-temperature, short-processing conditions and is thermally stable. Muses of oatrim is in foods like-such as frozen desserts, confectionery, baked goods, meat products, dairy products, and in cereals. The United States Department of Agriculture (USDA) has developed oatrim.

Z-trim- Z-trim stands for zero calorie trim. Z-trim was again developed by United States Department of Agriculture (USDA) to blend with oatrim. The high cellulose portion of hulls of oats, peas, soybeans, rice, or bran from corn or wheat are made from the indigestible insoluble fibres which are known as Z-trim. For processing, the hulls and brans are broken down into cellular fragments and purified, then dried and finally milled into a powder, to use this powder it is rehydrated into gel. Major applications of Z-trim include- it contributes to fibre and providing moistness, density, and smoothness to several foods, such as baked goods, cheese, and meat patties. The Z-trim's gel form is appropriate for frying hamburgers, etc. z-trim needs to follow completion of commercial process development, licensing as well as patenting for their commercial availability in the market.

Those substances that act like fat but are non-fat substances are known as fat replacers in foods. Fat-replacers are required to provide taste to the food products just like conventional fats, with minimal or no health risks and fewer calories. Fat replacers are majorly found in foodstuffs like- sour cream, cheese, baked goods, salad dressings, margarine, yogurt, sauces, and gravies. Current studies demonstrate that carbohydrate-based fat replacer and protein-based fat replacer do not have harmful effects on human health. Though, fat-replacers have lesser calorie value than the foods containing conventional fats, still some people makes up for this reduction as they are incline more towards fat-replacer foods.

REFERENCES

1. Agar (gelling agent, stabilizer, thickener). (2013). [Blog] juliedaniluk.com. Available at: https://www.juliedaniluk.com/13-agar-gelling-agent-stabilizer-thickener/view-details.html [Accessed 29 Jul. 2017].
2. Faia.org.uk. (2012). Food texture. [online] Available at: http://www.faia.org.uk/gelling-agents-thickeners-stabilisers/ [Accessed 29 Jul. 2017].
3. Healthline. (2005). Should You Remove Carrageenan from Your Diet? [online] Available at: http://www.healthline.com/health/food-nutrition/carrageenan#overview1 [Accessed 29 Jul. 2017].
4. Martinko, K. (2015). Emulsifiers in packaged foods are wreaking havoc with our digestive tracts. [online] TreeHugger. Available at: https://www.treehugger.com/health/emulsifiers-are-wreaking-havoc-our-digestive-tracts.html [Accessed 29 Jul. 2017].
5. Moncel, B. (2016). Where Does Gelatin Come From? [online] The Spruce. Available at: https://www.thespruce.com/what-is-gelatin-1328467 [Accessed 29 Jul. 2017].
6. Oulton, R. (2004). Pectin. [online] CooksInfo.com. Available at: http://www.cooksinfo.com/pectin [Accessed 29 Jul. 2017].
7. Panlasang Pinoy. Food High in Lecithin. [online] Available at: http://panlasangpinoy.com/2013/10/08/food-high-in-lecithin/ [Accessed 29 Jul. 2017].

CHAPTER 9

MAJOR FEED ADDITIVES

"Life is too short not to eat raw and it's even shorter if you don't."

—Marie Sarantakis

Feed additives are the food supplements that are given to the farm animals that are unable to get enough nutrients from the natural means. They include antibiotics, vitamins, minerals, amino acids etc. Feed additives are directly adjoined with the animal's diet in order to improve their value for human consumption. This chapter will drive us through different types of additives that are used in animal nutrition, some of them are antibiotics, beta- agonists, prebiotics, probiotics and somatotrophins.

OVERVIEW

Feed additives are the food supplements that are given to the farm animals that don't get enough nutrients from the regular means and includes antibiotics, vitamins, minerals, amino acids etc. The diet of humans and animals contain a broad diversity of additives. The additives in the poultry diets, however, are primarily comprised to improve the efficiency or efficacy of the bird's growth or laying capacity of bird, improve feed utilisationutilization and in prevention of disease. The additives that are used in feed should first be approved for use and then they must be used as per the direction given with respect to duration of feeding and inclusion levels. Feed additives are specific for the age and type of birds.

The guidelines for the usage of additives in feed are maintained by Australian government department of agriculture, fisheries and forestry. Some of the common feed additives that are used in poultry diets are- antimicrobials, emulsifiers, binders, antioxidants, pH control agents and various enzymes.

Some other additives like artificial and nutritive sweeteners, lubricants etc. are also used. There are dozens of companies which manufacture and distribute specific additives which each class of additives. Again, the ingredients and additives that are added must be noted on the label with their use and inclusion level meeting the standards as defined by the law.

In some cases, additives are directly adjoined with the animal's diet in order to improve their value for human consumption, but in most cases, this is accomplished by the use of natural ingredients which contain significantly elevated levels of these nutrients that are deposited straight into meat and eggs. Livestock farmers predominantly uses eight major categories of feed additives named- acidifiers, amino acids, antibiotics, antioxidants, binders, enzymes, minerals and vitamins.

ANTIBIOTICS AS FEED ADDITIVES

Beta-agonists

The synthetic compounds which mimic some of the impacts of normally happening mixes by authoritative to beta-receptors on the surface of cells inside the muscle, fat and different tissues of creatures are known as Beta-adrenergic receptor agonist which in short or are popularly termed as beta-agonist. Beta-agonists are utilized as a part of human drug for the treatment of conditions, for example, asthma. Be that as it may, different sorts of beta-agonists are utilized as a part of domesticated animal's generation to improve development and change body structure. These are the growth promoters highly available for the use in cattle.

Item	Opataflexx™	Zilmax™
Active ingredient	ractopamine hydrochloride	zilpaterol hydrochloride
Duration of feeding,d	28-42 days	20-40 days
Optimal feeding duration,d	28-35 days	20 days
Withdrawal time, d	None	3 days
Average daily gain, lbs/d	Incease from 9-21%	Increase from 14-40%
Feed efficiency (Feed:Gain)	Imprpve 9-21%	Improve 14-40%
Dressing percent	Minimal impact	1 to 2 percentage unit increase
Hot carcass weight, lbs	10 to 025 lbs increase	13 to 28 lbs increase
Ribeye area, sq.in	Up to 0.5 sq. in. increase	From 0.3 to 1.0 sq. in. increase
Marbling	Minimal impact	None to up to 25 degree reduction in marbling score
Tenderness	Minimal impact	None to reduction in 14 day Warner-Bratzler shear force up to 1.7 kg

Figure 9.1: Characteristics of beta agonists approved for beef cattle.

Source: http://fyi.uwex.edu/wbic/files/2010/11/Beta-Agonists-Factsheet.pdf

Fro Eelanc Aanima Hhealth a product is found known as Optaflexx. Optaflexx has an active ingredient present in it i.e. ractopamine hydrochloride, it is that ingredient which has been labelled only for the use of swine. That particular ingredient is paylean. Another ingredient named as Zilmax is available from intervet, Inc. Zilpaterol hydrochloride is the active ingredient present in it.

Difference Between Zilmax and Optaflexx

Generally, if we see, Zilmax is a more intense beta agonist than Optaflexx, though most examinations have watched more prominent red meat yield with encouraging Optaflexx verses Zilmax. The contrasts between these two items in affirmed name guidelines and execution of cattles are sketched out in the above table. Although no negative impact has ever been sited on the animals but, the cattles who are poorly legged and are straight fronted than the added muscles in these products can be a problem for them.

Use of Beta-agonist in cattle feed

Zilmax are the food additives mostly used by the cattle feeders to build the rate of weight pick up without extra increase in nourishment to make feed proficiency more noteworthy. On August 7, 2013, a leading company of beef processing announced that they would no longer purchase cattle who have been injected Zilmax. The company's name was Tyson Foods and they did this because of their concern or the welfare of the animals.

According to Tyson there were occurrences of steers experiencing stress and weakness after transportation to butcher plants. Further, it was also noticed the issue was not one of sustenance wellbeing, but rather of creature welfare. With an eye on their clients and their view of nourishment quality, regardless of whether affirmed or not, Tyson made an immense stride in reducing those worries. On August 16, 2013 Merck and Co., the creators of Zilmax, suspended the offer of the item in the US and Canada.

Merck and Co the creators of Zilmax, selected creature welfare master Dr. Temple Grandin to assist counsel with a survey of the item. In a meeting with Rueters on August 22, Dr. Grandin showed there have been episodes of firmness, soreness, and warmth worry in pressing plants since the utilization of beta-agonists started. In situations where issues were seen, Grandin said the greater part of the steers would be fine at the feedlot, yet 50% of them would demonstrate some level of stress or faltering indications after landing in the pressing plant. While these conditions were not appeared in the wellbeing endorsement process for Zilmax, such welfare occurrences and activities by Tyson Foods could make the item inaccessible for a significant timeframe as the assessment procedure proceeds.

Beta-agonists are a class of non-hormonal mixes bolstered to steers. Their method of activity is to tie to receptors on fat cells in the creatures' body and divert and lessen the digestion of fat. Thus, less fat is created and less fat is put away in the body. In the meanwhile, the mixes tie to receptors on muscle cells and divert and increment the extent of muscle filaments. Muscle fibre estimate replaces a portion of the weight regularly found from fat and the aggregate remains contains a higher rate fit muscle. These activities lessen the vitality provided by the encourage to create weight pick up. With more weight created by a similar level of bolster consumption sustain effectiveness is expanded.

The items experience a broad endorsement process by the Food and Drug Administration before they can be utilized as a part of domesticated animal's bolsters. There are two mixes accessible to cattle feeders: Zilmax (zilpaterol) made by Merck and Co. and Optaflexx (ractopamine) made by Eli Lilly Co. The two items are bolstered at low levels (200 mg/head/day for Optaflexx and 60-90 mg/head/day of Zilmax) for a brief span just before butcher (25-35 days). A current report recommended 60% to 80%

of feedlot steers in the US are encouraged a beta-agonist (K. Micik, 2013.) The Encyclopaedia of Meat Science (altered by Dikeman and Devine; 2004) detailed Optaflexx expanded normal day by day pick up by 15-25% with no extra sustain consumption. Marginally higher outcomes are appeared for Zilmax. Neither one of the products has demonstrated any impact on meat quality at acknowledged nourishing rates, and lean yield of the corpse was expanded. Littler steers feedlots from time to time utilize beta-agonists. At the point when pens of dairy cattle are being arranged week after week or bi-week after week available to be purchased, it ends up noticeably difficult to successfully utilize these items in view of the characterized bolstering period preceding butcher. Sometimes, be that as it may, they might be utilized as a part of pens of littler confined yearlings or cows to expand cadaver measure from the extra weight pick up and(or) diminish fat and enhance yield review. This serves to abstain from showcasing rebates for little cadavers (under 600 lbs) or YG4 or YG5 bodies. Sustaining the added substance and showcasing must even now be done on a pen premise.

Since such an extensive number of dairy cattle are sustained beta-agonists, the loss of one of the items can significantly affect beef markets. The report from Micik (2013) cites DTN Market Livestock Analyst John Harrington as saying a lessening of 10-15 lbs. off of all meat remains because of the diminishment being used of beta-agonists is 0.5 billion pounds of hamburger, or around 1% to 1.5% of generation. Regularly, this outcome would raise the cost of live steers by $3 to $4 as per Harrington. Thus, different elements of the meat showcase and monetary profiles for dairy cattle nourishing could be affected. One illustration is more corn expected to complete steers which raises the opposition for other domesticated animal's employments. Cattle feeders will have the choice of moving to Optaflexx in them apportions since this item is as yet accessible. Be that as it may, there may even now be a little diminishment in weight for every corpse. Corn costs that pattern bring down urge dairy cattle feeders to sustain steers longer and increment corpse weights, and this outcome could counterbalance some effect of not utilizing beta-agonists. Innovation will drive the change of meat generation and keeping sustenance costs low now and later on. Difficulties in the utilization of food additives, drugs, implants and other innovation will moderate that change. In any case, clients require a positive view of the product they purchase, and it is important to viably respond to all their concerned queries.

Impact of Beta-agonist on carcass traits and the quality of meat

Beta agonists provides more vitality to muscle than fat deposition. Many research contemplates have announced expanded carcass weight and ribeye territory at a comparable fat thickness. Whenever cattles are nourished with Optaflexx, then according to a report its impact is negligible and the quality and the tenderness of the beef is much better than those cattles who are bolstered with Zilmax. A few examinations have announced negligible effects on quality review and delicacy, while other have revealed slight decreases in marbling resulting in less dairy cattle reviewing USDA Choice and increment in durability. The extent of reaction in these investigations could have been impacted by organic sort of steers. Furthermore, more noteworthy reactions have been seen in dairy cattle sustained in little explore pens verses in substantial business feedlot pens.

According to various reports, beta-agonists are liable and effective only for 28-35 days. After this time duration, the performance level of the cattles returns back at the same level where they stood before the intake of this product. The cattles loses their effectiveness to work and perform as their body gets adaptive to the active ingredient during that particular time-period. And not to forget if the cattle feeder injects more than required quantity of the product than it would be considered as an illegal approach towards the cattle. These products are not supposed to be use for the purpose of animal breeding. It is an offence and the person found doing this would be charged with legal actions.

Antibiotics

When human beings get infected by any bacteria or any virus which causes them illness, they are then provided with antibiotics to reduce the impact and effect of those bacteria. Similarly, when an animal is diagnosed with any sort of illness then not just that animal but the entire batch of animals are treated with antibiotics so that it does not get transmitted to any other animal. They are chemotherapeutic agents used to cure the infectious disease in humans, animals and plants through clinical management. According to a report, US alone uses approximately 24.6 million pounds of antibiotics every year for animals and their treatment. It is believed that micro-organisms present in the cattle feed consumes a substantial amount of nutrients in the feed. This could lead to the formation and production of toxins, which can further have adverse effects on the cattles.

The development advancing impact of anti-toxins may originate from their capacity to stifle these unsafe living beings. It is likewise proposed that creatures raised in the unhygienic conditions dependably bear some idle diseases, which trigger a course of occasions in their immune system. Cytokines created in the process prompt the arrival of some catabolic hormones which cause wastage of muscles. Anti-infection agents assuage the creatures of the need to deliver cytokines by smothering the causative specialists of diseases.

Use of antibiotics in cattle feed

It has been almost 50 years that antibiotics are used in animal feed. They are used not only as an anti-microbial agent, but also as a growth-promoting agent and for the purpose of improving the performance of the animals. Tetracyclines, penicillin, streptomycin and bactrican are some of the antibiotics which are used as feed additives in livestock and poultry feed. But at present the highly recommended antibiotics to be used in livestock of the cattles are: - chlortetracycline, procaine penicillin, oxytetracycline, tylosin, bacitracin, neomycin sulfate, streptomycin, erythromycin, linomycin, oleandomycin, virginamycin, and bambermycins. Almost all these antibiotics are of microbial origin but there are other chemically synthesized antimicrobial agents that are also many a time used in animal feeds. These include three major classes of compounds: arsenical, nito-furan, and sulfa compounds. Arsenical compounds include arsanilic acid, 3-nitro-4-hydroxy phenylarsonic acid, and sodium arsanilate; nitro-furan compounds include furazolidone and nitro-furazone; sulfamethazine, sulfathiazole, and sulfaquinoxaline. Other chemicals are also used as antiprotozoal agents to prevent coccidiosis and histomaniasis in chickens and turkeys. Antibiotics are used regularly in animal feed at a rate of 2 to 50 grams per ton for improving the performance in the animals. The reasons include a more efficient conversion of feed to animal products, an increased growth rate and a lower morbidity/mortality rate in general. If there is any specific disease which can spread at a very high rate than the level of antibiotics gets increased to 50-200 grams/ton or even more.

The levels are also increased in times of stress. After the threat of the spread of the disease reduces the amount of the antibiotics injected in the feed is reduced.

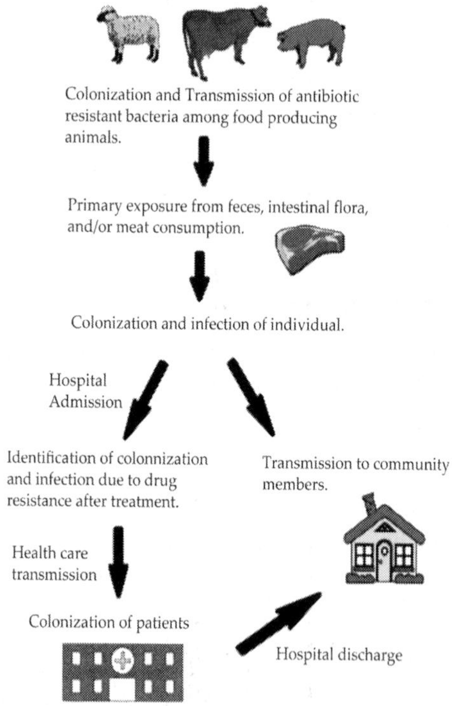

Figure 9.2: Flowchart showing the effect of antibiotics on animal feed

Source: http://www1.udel.edu/chem/C465/senior/fall97/feed/present.html

Importance of antibiotics in cattle feeds

Antibiotics used as a feed additive are of great importance. Pigs supplemented with antibiotics, to achieve a desired level of growth in their feed requires almost 10–15% less feed. The feeds are expensive and leads to a huge amount of investment in rearing the animals. Therefore, when antibiotics are used as feed additives in the feed of cattles than it cuts down the expenditure, not at a big level but at a minimal level. Antibiotics added to the feed improves the quality of feed as it gets converted to a feed to animal product. It has been noticed and also reported that the daily growth rate of animals subsisting on antibiotic-supplemented food improves by 1–10% in comparison to the feed provided to them without antibiotic. According to Hughes and Heritage, in the report flashed in the year 2002 stated that the meat which we purchase, it is fed with antibiotics than its quality is much better constituting of higher amount of protein and less amount of fat compared to that obtained from animals not supplemented with antibiotics.

According to Gustafson and Bowen in their report in 1997, Utilization of antibiotic medication and penicillin in chicken sustain prompted a critical change in the creation of eggs and hatchability other than encourage effectiveness. Wellbeing of the domesticated animals bolstered with anti-microbial blended sustenance is additionally uniquely moved forward. Following expansion of chlortetracycline and sulfamethazine to the sustain, the rate of ox-like respiratory illness dreariness, the rate of backslides and mortality and furthermore the rate of creatures determined to have constant respiratory ailment were observed to be essentially diminished (Gallo and Berg, 1995). Advantages as far as the rate and proficiency in the increase in the weight, diminish in mortality and grimness and decrease in the event of subclinical illnesses, were seen during the use of anti-infection agents amid all periods of development of pigs (Cromwell, 2002). The unfavorable of inflammation and pro-inflammatory mediators go between in creatures like, no growth or downfall in the growth, bolster consumption, reproduction, production of milk, and metabolic wellbeing are not hidden from anyone. The mitigating capability of anti-microbials (especially macrolides) gives a balanced premise of their gainful impacts which is free of their antimicrobial impact (Buret, 2010). Therefore, we can say that antibiotics plays a crucial role in profitable and efficient production of the livestock.

Risk of antibiotics in animal Feed

Animals retain the strains of bacteria which are resistant to antibiotics, this happens after the animals have been fed antibiotics. bacteria proliferate in the animal. Colonization of antibiotic resistant bacteria gets formulated, due to the interaction and the resistance of bacteria that get transmitted to other animals. The bacteria flourish in the intestinal flora of the animal, as well as, in the muscle. As a result, the faeces of the animal often contain the resistant bacteria.

These bacteria get transmitted to humans from animals through various means of transformation. The primary exposure of humans to resistant bacteria occurs in farms and butcher houses. The faeces of the animals which contain the bacteria of the animals on farms gets cleaned by the human being before they cut the animal. During the cleaning process, humans may get bacteria on their body and hands. If the body or hands are not properly cleaned, the bacteria could be ingested by the person. Likewise, in slaughterhouses, during slaughter, the intestine is severed. Resistant bacteria are exposed to slaughterhouse workers, which could get the bacteria on their bodies and hands. Transmission occurs when the bacteria is ingested. Along with the previous sources of contamination, humans get infected to the disease by consuming meat of the animal which had already had the disease. Even though cooking reduces the survival of the bacteria, some may still survive and infect the human. For example, in the year 1983, 18 people in four midwestern states developed multi-drug resistant Salmonella food poisoning after they had beef from cows fed antibiotics. After initial transmission and infection to humans, the transmission to other humans has a couple paths. Transmission can occur through numerous mediums (airborne, physical contact, and organic liquids) of human contact in the group. A tainted individual may likewise be admitted to a healing center for treatment. Treatment may not work in tranquilize safe microscopic organisms, in this manner, distinguishing a medication safe contamination. Microbes is transmitted to different patients by means of the doctor's facility condition or human services workers hands. After transmission, the microbes will colonize in a few of the patients. Colonization in different patients with other safe microscopic organisms can create microorganisms with multi-tranquilize resistance. Once the patients recuperate, they are released into the group. These patients could possibly taint a few group individuals. Numerous disease could conceivably create a supergerm which is impervious to many medications because of resistance sharing between microbes.

Probiotics and Prebiotics

Probiotics are those live microorganisms that depends upon the direct feed in the natural environment and leave a positive influence on the behavior of the host such as fungi and yeast. Similarly, prebiotics are a feed for microflora which can alter the composition of microflora by specifically providing appropriate nutrients for the favorable bacteria to grow. Thus, when combined both the probiotics and prebiotics together they are called as synbiotics to increase the production of flocks and promote the expansion of feed additives.

Probiotics

Probiotics refers to the task of sustaining an appropriate balance of helpful microorganisms in the intestine of birds and animals that serves as important measure for the efficient transformation of feed into better productivity and growth. These microbial bacteria prevent the formation of harmful microorganisms that produces dangerous toxins. These prebiotics have many beneficial effects and some are listed below:

At first, the notion of competitive exclusion of microbials was tested when it was seen that via oral intake of fowl intestinal bacteria decreased the rate of *Salmonella* infection in the little chicks. Hence, this fine characteristic of probiotics is rapidly used in the production of animal and poultry to reduce the dangerous impact of pathogenic bacteria.

There are few probiotic microorganisms like *Lactobacillus* that suppress the rate of growth of some bacteria like *Salmonella* by itself producing multiple bactericidal products which then further helps to inhibit the rate of intestinal level of pH.

To make out the best use of available substrates it is necessary for the probiotic strains to evolve in the surroundings of gut such that it can use the available nutrients that might be available to rest of the harmful microorganisms.

Probiotics helps to increase the rate of appetite which helps further in enhancing of digestion of essential nutrients in the body of host. They are very helpful in the metabolic process of minerals and digesting the vitamins.

The excessive intake of ammonia can cause damage to the epithelial lining in the tract of digestive system which may lead to the problems such as keratoconjunctivitis in the farm units. So, to reduce the impact of this bacteria of ammonia the probiotics are used as the opponent to suppress the formation of bacteria in the intestine of birds.

Probiotics serves as a very useful component to develop the immune resistance in neonates by using the immune-stimulatory action by multiplying the production of immunoglobulins and by stimulation of cell activities which provides the local immunity to the host.

Probiotics in livestock and poultry production systems

There is a wide scope of probiotics in the field of poultry and livestock. It has been observed that if there is any activity which disturbs the ratio of gut microflora due to the presence of harmful microorganisms, will in turn lead to the reduction in the performance of animals and birds will ultimately cause the financial loss in the business of livestock and poultry. Earlier, to prevent the occurrence the such type of situations, the use of antibiotics was used as a feed additive. But, when this was used in the long term it has created the concerns regarding the health of public where they were indirectly intaking the antibiotics in the form of meat or milk. Thus, by keeping the track of the past and its harmful effect on the public as a whole, the new concept of probiotics was emerged which can be utilized for the feed to maintain the balance of microflora which can improve their productivity and increase the rate of metabolism to maintain good health. There are various uses of probiotics such as they help to improve the digestion of nutrients, increases the rate of productivity, excludes the harmful organisms that can produces diseases, prevents infection that may cause diarrhea, reduction in stress after the process of vaccination and finally it helps to stimulate the responses towards immunity. Through the religious use of probiotics, it leaves an influential impact on the immune system that incorporates the formation of natural interferons, immunoglobins and immunity stimulated by cell mediation. In the area of poultry, probiotics are really helpful as they enhance the rate of fertility of eggs, increases the quality of production of egg and lessens the rate of mortality besides above given advantages.

Prebiotics

Oligosaccharides are the component of carbohydrates that are used to trigger the selective number of bacteria to operate the gut ecosystem. Thus, the substrates which are used for this process are known as prebiotics which are used for the enrichment of useful microorganisms those who are capable of replacing the harmful bacteria from the gut microflora. Prebiotics are referred as an important constituent that are not digestible for the parent body rather it is fermented occasionally by other useful microflora in gut. They are well equipped with the capability to transform the equal ratio of bacteria in a healthy way to the parent body.

Prebiotics, is a new concept in the study of growth promoters that has a major concern with the fermentation done on the selective intestinal flora. When prebiotics are assisted with probiotics they have the aligned effect on the feed ingredient, immunity stimulation and offsetting effect on the toxins which results in the lower pH value which inhibits the generation of pathogenic microorganisms. Over the years, it has been researched that the bacteria of lactic acid are used to increase the resilience against any disease which can be enriched into the tract of intestines by intaking specific types of carbohydrates. Along with this, there are indigestible oligosaccharides that are used for the stimulation of various types of minerals and can be helpful for the mineralization of bone. It has been stated that the prebiotics enhances the productivity of necessary minerals like calcium, zinc and iron in the parent body.

Recent trends

The innovation of advanced molecular techniques in biology and adaptation of various modern tools have paved the path for enriched amenities in the food research. In many of the international institutes, the differentiation and identification of strain probiotics in feeding stuff is considered. There exists another feasible methods to select and distribute probiotics from other microbes such as RAPD-PCR.

It is likely to create entirely new probiotic by exploiting the genetic engineering methods. The genetically modified probiotics have the probability to revolutionize alimentary health if they are designed wisely and cautiously. Genetic engineering methods, such as Recombinant DNA technique has been used for the modification of lactic acid bacteria to produce novel varieties or modification of these bacteria for use in oral immune-therapeutic applications.

REFERENCES

1. Bernardeau, M., Guguen, M. and Vernoux, J. (2006). Beneficial lactobacilli in food and feed: long-term use, biodiversity and proposals for specific and realistic safety assessments. FEMS Microbiology Reviews, 30(4), pp. 487-513.

2. Bevenovo. Feed Additive Classification - Technical Data - Bevenovo. [online] Available at: http://www.bevenovo.com/TechnicalData/66.html [Accessed 29 Jul. 2017].
3. Dhama, K. and Tomar, S. (2008). Beneficial Effects of Probiotics and Prebiotics in Livestock and Poultry: The Current Perspectives. [pdf] Available at: https://www.researchgate.net/publication/229596840_Beneficial_effects_of_probiotics_and_prebiotics_in_livestock_and_poultry_the_current_perspectives [Accessed 29 Jul. 2017].
4. Nawaz (2017). feed additives, their classification, mode of action & use in ruminants. [online] Slideshare.net. Available at: https://www.slideshare.net/vicky14381/feed-aditives-their-classification-mode-of-action-use-in-ruminants [Accessed 29 Jul. 2017].

CHAPTER 10

RECENT ADVANCEMENTS AND DEVELOPMENTS IN BIOTECHNOLOGY OF FOOD ADDITIVES

"You cannot achieve environmental security and human development without addressing the basic issues of health and nutrition."

-Gro Harlem Brundtland

Biotechnology is directly linked to food handling as most of the developing countries are making use of microbial inoculants to improve properties, such as- taste, smell, shelf life, surface, and dietary estimation of foods. This chapter will drive us through the recent developments and applications of modern biotechnical tools and methods in food processing. This chapter also includes the use of biotechnology in food safety and quality improvement, including the food safety issues and concerns, advancements in microbial genetics, mycotoxin detection, etc.

OVERVIEW

Preparing food makes use of different unit operations and innovations to change over generally cumbersome, perishable, and unappetizing crude materials into more helpful rack steady and agreeable foods or consumable drinks. Processing adds to food security by limiting waste and misfortunes in the natural way of life and by expanding food accessibility and marketability. Food is additionally handled to enhance its quality and wellbeing. Food security is a logical train that gives affirmation that food will not make hurt the shopper when it is readied or potentially eaten by its expected use.

Biotechnology as connected to food handling in most creating nations makes use of microbial inoculants to improve properties, for example, the taste, smell, time span of usability, surface, and dietary estimation of foods. The procedure whereby small-scale life forms and their compounds achieve desirable changes in food materials is known as fermentation. Fermentation processing is additionally broadly connected in the creation of microbial societies, compounds, flavourflavors, aromas, food added substances and a scope of other high esteem included items.

APPLICATION OF TRADITIONAL AND NEW BIOTECHNOLOGIES IN FOOD PROCESSING

Methods of microbial inoculation in food fermentations- The fermentation bioprocess is the major biotechnological application in food preparing. It is regularly one stage in a succession of food preparing operations, which may incorporate cleaning, measure diminishment, drenching and cooking. Fermentation bioprocessing makes utilization of microbial inoculants for improving properties, for example, the taste, fragrance, time span of usability, wellbeing, surface and dietary estimation of foods. Microorganisms related with the crude food material and the handling condition fill in as inoculants in unconstrained fermentations while inoculants containing high groupings of live small-scale living beings, alluded to as starter societies, are utilized to start and quicken the rate of fermentation procedures in non-unconstrained or controlled fermentation procedures. Microbial starter societies shift generally in quality and purity.

Unconstrained inoculation of fermentation processes

At the rural households and villages fermented foods are produced using Unconstrained methods of inoculation. Unconstrained fermentations are to a great extent uncontrolled. A characteristic determination process, be that as it may, develops in large portions of these procedures which in the end brings about the power of a specific sort or gathering of group of micro-organisms in the fermentation medium. Real impediments of unconstrained fermentation procedures incorporate their wastefulness, low yields of item and variable item quality. While unconstrained fermentations by and large improve the security of foods attributable to a lessening of pH, and through detoxification, at times there are wellbeing concerns identifying with the bacterial pathogens related with the crude material or unhygienic works on amid handling. A larger part of African food fermentation procedures makes utilization of unconstrained vaccination.

Appropriate starter cultures as inoculants of fermentation processes

Appropriate starter societies are generally connected as inoculants over the fermented food area, from the household to industrial level. These starter societies are for the most part created utilizing a backslopping procedure which makes utilization of tests of a past cluster of a fermented item as inoculants. Appropriate Proper starter societies are broadly connected in the creation of fermented fish sauces and fermented vegetables in Asia and in oat or grain fermentations in African and Latin American nations. The immunization belt utilized as a part of conventional fermentations in West Africa fills in as a bearer of unclear fermenting micro-organisms, and is one case of an appropriate starter culture. It for the most part comprises of a woven fiber or tangle or a bit of wood or woven wipe, immersed with "high"- quality result of a formerly fermented bunch. It is inundated into another bunch, keeping in mind the end goal to fill in as an inoculant. The immunization belt is utilized as a part of the generation of the indigenous fermented porridges,"uji" and "mawe", and also in the creation of the Ghanaian lager, "pito". Iku, likewise alluded to as iru, is yet another case of an appropriate starter culture delivered by backslopping. This starter culture is created from concentrated fermented dawadawa (a fermented vegetable item), blended with ground unfermented vegetables, vegetables, for example, pepper, and oats, for example, ground maize. It is put away in a dried frame and is utilized as an inoculant in dawadawa fermentations in West Africa. A scope of appropriate starter societies, either in a granular shape or as a squeezed cake is utilized crosswise over Asian nations as fermenting inoculants. These customary form starters are for the most part alluded to by different names, for example, marcha or murcha in India, ragi in Indonesia, bubod in the Philippines, nuruk in Korea, koji in Japan, ragi in Malaysia and Loog-throb in Thailand. They for the most part comprise of a blend of molds become under non-sterile conditions.

Defined starter cultures as inoculants of fermentation processes

Few defined starter cultures have been created in research center for use as inoculants in business fermentation forms. "Defined starter cultures" comprise of single or blended strains of microorganisms. They may consolidate aide culture arrangements that serve a food wellbeing and additive capacity. Aide societies don't really create fermentation acids or alter surface or flavor, however are incorporated into the characterized culture inferable from their capacity to hinder pathogenic or decay living beings. Their inhibitory movement is because of the generation of one or a few substances, for example, hydrogen peroxide, natural acids, diacetyl and bacteriocins.

Defined cultures are created by immaculate culture support and engendering under aseptic conditions. They are for the most part showcased in a fluid or powdered shape or else as a squeezed cake. Loog-string, a characterized culture promoted in Thailand as a squeezed rice cake, comprises of Saccharomyces cerevisiae, Aspergillus oryzae or Rhizopus sp. what's more, Mucor. Loog- string has a time span of usability of 2–3 days at encompassing temperature and 5–7 days under refrigerated conditions. Ragi societies are monetarily created by the Malaysian Agricultural Research and Development Institute by blending a culture inoculum which by and large comprises of Rhizopus Oligosporus with saturated sterile rice flour, and brooding it at surrounding temperature for four days. This starter has a time frame of realistic usability of two weeks under refrigerated conditions. It is generally utilized as an inoculant in the creation of customary Malaysian fermented foods. Ragi-sort starter societies for the creation of a scope of fermented Indonesian items, for example, oncom, tape and tempeh are at present promoted by means of the internet.

Defined starter cultures are additionally broadly utilized as a part of the business generation of dairy items, for example, yogurt, kefir, cheeses and mixed drinks. A large number of these cultures are customized to deliver particular surfaces and flavors. In light of developing customer enthusiasm for accomplishing wellbeing through eating regimen, numerous yogurt cultures additionally incorporate probiotic strains. Probiotics are as of now delivered in India for use as food additives, dietary supplements and for use in animal feed.

Defined starter cultures developed using the diagnostic tools of advanced biotechnologies

The utilization of DNA-based indicative strategies for strain separation can take into consideration the fitting of starter cultures to yield items with particular flavors particular flavors and additionally surfaces. Irregular enhanced polymorphic DNA (RAPD) methods have been connected in, for instance, Thailand, in the atomic writing of bacterial strains and relating the discoveries of these examinations to season improvement amid the creation of the fermented pork wiener, nham. The consequences of these examinations prompted the improvement of three distinctive characterized starter cultures which are as of now utilized for the business creation of items having diverse flavor attributes.

GM starter cultures

To date, no business GM micro-organisms that would be devoured as living life forms exist. Results of modern GM maker life forms are, be that as it may, generally utilized as a part of nourishment preparing and no real wellbeing concerns have been raised against them. Rennet which is generally utilized as a starter in cheddar creation over the globe is delivered utilizing GM microbes. Thailand at present makes utilization of GM Escherichia coli as an inoculant in lysine creation. Numerous mechanically imperative proteins, for example, α-amylase, gluco-amylase, lipase and pectinase and bio-based fine chemicals, for example, lactic corrosive, amino acids, anti-infection agents, nucleic corrosive, and polysaccharides, are delivered in China utilizing GM starter societies. Different nations which as of now create chemicals utilizing recombinant smaller scale life forms incorporate Cuba, Brazil, India, and Argentina.

Chemicals, amino acids, vitamins, natural acids, polyunsaturated unsaturated fats and certain intricate starches and enhancing specialists utilized as a part of nourishment definitions are as of now created utilizing GM micro-organisms.

APPLICATION OF BIOTECHNOLOGY IN FOOD SAFETY AND IN QUALITY IMPROVEMENT

Food safety issues and concerns in food fermentation processing

Microbial activity assumes a focal part in food fermentation forms, bringing about desirable properties, for example, changes in time frame of realistic usability and quality characteristics, for example, surface and flavor. Pathogenic organisms are, be that as it may, of prime worry in fermented foods. Hostile to dietary factors, for example, phytates, tannins, protein inhibitors, lectins, saponins, oligosaccharides and cyanogenic glucosides are normally happening parts of crude materials generally utilized as a part of food fermentations. Defilement of the fermentation procedure can represent a noteworthy wellbeing hazard in the last fermented item. Systems for distinguishing and checking the nearness of concoction (pesticide buildups, overwhelming metals, follow components) and biochemical (aflatoxins) dangers in fermented foods are, in this way, a basic need. Besides, with developing buyer enthusiasm for the assurance traits of the items that they expend, and the premium as of now being put on quality connected to land starting point, the traceability of foods with chose properties is of expanding significance.

Advances in microbial genetics

As of late, the hereditary portrayal of micro-organisms has progressed at a fast pace with exponential development in the accumulation of genome arrangement data, high-throughput investigation of communicated items i.e., transcripts and proteins and the use of bioinformatics which permits high throughput near genomic approaches that give bits of knowledge to advance practical examinations. Genome grouping data, combined with the help of exceptionally progressed atomic methods, have enabled researchers to set up components of different host-cautious pathogen counter-guarded techniques and have furnished industry with instruments for creating methodologies to plan sound and safe food by enhancing the impact of probiotic microorganisms, the outline of starter culture microscopic organisms and utilitarian properties for use in food preparing. Portrayal of the genomes of lactic corrosive probiotics has, for instance, shed light on the connection of pathogens with lactic corrosive microscopic organisms. Nucleotide successions of the genomes of numerous essential food organisms have as of late turned out to be accessible. Saccharomyces cerevisiae was the principal food microorganism for which a total genome succession was described. This was trailed by genome sequencing of the related yeast, Kluyveromyces lactis and also filamentous parasites fungus which are real compound makers and have noteworthy applications in the food preparing industry.

Detection of pathogens

The rapid identification of pathogens and other microbial contaminants in food is basic to evaluate the wellbeing of food items. Customary strategies to identify food borne microorganisms frequently depend on tedious development in culture media, trailed by disengagement, biochemical recognizable proof, and now and again serology. Late mechanical advances have enhanced the proficiency, specificity and affectability of identifying micro-organisms. To Identification identify advancements microorganisms utilize the polymerase chain response reaction (PCR) is examineused. Short pieces of DNA (tests) or groundworks are hybridized to a particular grouping or format, which is in this manner enzymatically enhanced by the Taq polymerase chemical utilizing a thermocycler. In principle, a solitary duplicate of DNA can be opened up a million-crease in under 2 hours with the utilization of PCR methods; henceforth, the capability of PCR to dispose of or significantly lessen the requirement for cultural advancement. The hereditary portrayal of genome grouping data has additionally encouraged the recognizable proof of destructiveness nucleotide arrangements for use as sub-atomic markers in pathogen discovery. Multiplex continuous PCR strategies are currently accessible to recognize the E. coli O157:H7 serogroup. PCR-based recognizable proof strategies are likewise accessible for Vibrio cholera and for significant food related organisms, for example, Campylobacter jejuni, C. coli, Yersinia enterocolitica, Staphylococcus aureus.

Mycotoxin detection

The problem of mycotoxin defilement in food including fermented foods is a worldwide concern. Mycotoxin pollution is especially pervasive in creating nations in tropical territories, for example, in South Asia and Africa. High-performance liquid chromatography Superior fluid chromatography (HPLC) and gas chromatography/mass spectrometry (GC/MS) are two of the most generally utilized strategies for the location and evaluation of mycotoxins in developing nations. These strategies, in any case, are tedious, hard to utilize and require research center offices. Immunoassays that are conservative being used, delicate and simple to utilize would enormously encourage the identification and quantization of mycotoxins. Various ELISA packs are currently economically accessible for the recognition of aflatoxins, deoxynivalenol, fumonisins, ochratoxins, and zearalenone.

Detection and identification of foods and food ingredients

The DNA-based ID code framework is dependent on polymorphisms at the nucleotide level for the separation of living organisms at the assortment and species levels. Right now, PCR-based strategies are utilized either with the end goal of identifying single nucleotide polymorphisms (SNPs) offering ascend to confinement piece length polymorphisms (RFLPs) or for distinguishing little grouping length polymorphisms (SSLPs) regularly known as Variable Number Tandem Repeats (VNTRs). These techniques encourage the distinguishing proof of interesting polymorphisms of an assortment of food wares and can be utilized as a part of the recognizable proof of their source or starting point. These one of a kind polymorphism are frequently alluded to as DNA barcodes. The DNA standardized identification is utilized for the ID of particular assortments in food recognition and in food traceability. The DNA standardized identification has been utilized for the distinguishing proof of numerous items for send out in nations, for example, Thailand, China, Brazil, Cuba and Argentina, The DNA standardized tag of microsatellite markers has additionally been effectively utilized as a part of separating and recognizing fermented items, for example, premium wines, cheesescheeses, and wieners on the premise of their beginnings. Basmati rice assortments and olive cultivars utilized as a part of olive oil generation have additionally been separated.

A key issue in the sector where the application of biotechnologies could be useful

Emerging pathogens

The distinguishing proof of irresistible operators requires top of the line innovations which are not generally accessible in developing nations. Developing nations must, in this manner, look for help from nations with higher bore advancements so as to portray the irresistible specialists, set up reconnaissance and observing frameworks and create methodologies to contain the disease(s). Biotechnology can assume a key part in encouraging the portrayal of new developing pathogens. Customary social strategies for the recognition and specification of microbial pathogens are monotonous and require no less than 12–18 hours for the acknowledgment of results. At that point, the food items would have been disseminated to retailers or customers. Immunoassay analytic packs encourage close constant checking, affectability, adaptability, and usability. The rise of multi-anti-infection resistance characteristics is common in escalated cultivating in creating nations because of the mishandle of anti-toxins. The spread of multi-anti-infection safe microorganisms postures general wellbeing concerns, since pathogens showing such resistance would be hard to control with the utilization of presently accessible anti-infection agents. The quick discovery of these pathogens, with high affectability, is one method for observing and containing the spread of multiantibiotic safe characteristics. A key approach being utilized by some is the advancement of fondness biosensors with an anti-infection safe nucleotide succession as the location test.

Distinguishing alternatives for creating nations- It is imperative that nations perceive the capability of fermented food and organize activities to guarantee their security, quality and accessibility. In view of the stocktaking exercise in this archive, various particular choices can be recognized for creating nations to enable them to settle on educated choices with respect to selection of biotechnologies in food preparing and in food wellbeing for what's to come.

Administrative and approach issues

- Governments must be focused on securing purchaser wellbeing and interests, and to guaranteeing reasonable practices in the food area.
- There must be agreement at the most abnormal amounts of government on the significance of food security, and the arrangement of satisfactory assets for this reason.
- Government strategy that depends on a coordinated natural pecking order approach is science-based, straightforward and incorporates the cooperation of the considerable number of partners from ranch to table must be set up
- The significance of the local and worldwide measurements of the utilization of biotechnologies in food preparing and wellbeing must be perceived.
- Need must be concurred to advancing fermented food in the food security plans of nations.
- Governments should likewise give an empowering domain that is steady of the development and advancement of upstream fermentation procedures, for example, the generation of high-esteem fermented items, for example, proteins, useful food fixings and food added substances.

Global participation and harmonization

The association and usage of territorial and universal fora are basic prerequisites for the improvement of national authoritative capacity and execution and for the help of worldwide co-operation. Further, the setting up of managerial structures with obviously characterized parts, obligations and accountabilities could effectively administer prepared foods and security issues.

Biotechnology-based Standard Working Methodology (SOPs) for food wellbeing ought to likewise be archived for use in approved research facilities. 18

National food control databases for the efficient gathering, revealing and examination of food related information (food review, investigation, and so on.) with set directions and guidelines in view of sound science and as per universal proposals (Codex) are key prerequisites.

Awareness policy

While the utilization of fermented food is developing in fame among higher-salary customers on account of expanding enthusiasm for health through eating routine, the utilization of fermented food by bring down wage shoppers in many creating nations is seen to be a regressive practice.

Techniques ought to in this way be created for the spread of learning about food biotechnology and, especially, fermented food. Directed shopper training on the advantages of consuming fermented food items and on applying great practice in their creation is required.

Food biotechnology ought to be incorporated into instructive educational module to enhance the learning base in nations on the commitment of fermented foods to food and healthful security and to create attention to the developing business sector open doors for fermented food and high-esteem items got from fermentation procedures.

Data sharing

Access to specific specialized data on biotechnology and biotechnological improvements in the food preparing division are basic and important information sources and emotionally supportive networks for controlling and arranging the examination motivation of nations. The important data frameworks ought to thusly be produced to encourage quick access to data on biotechnological improvements crosswise over both the created and the creating scene.

Enactment and policy on advances

Aptitude in enactment and innovation authorizing, and additionally learning about how to support advancement and transform it into business wanders, are basic prerequisites for developing nations. Effective innovation exchange requires these components and a situation that is helpful for development. Government arrangement in creating nations ought to in this way organize innovation exchange that makes new business wanders, an approach that requires government bolster, for example, impose impetuses and framework speculation.

Intellectual property rights (IPR)

A hefty portion of the customary fermentation procedures connected in creating nations depend on conventional learning. Upgraded specialized and logical data is required keeping in mind the end goal to assert responsibility for conventional information of the specialty of indigenous fermented foods. Absence of specialized learning has brought about the inability to understand the advantages of the industrialization of indigenous fermented foods by people who are the legitimate proprietors of the innovation.

More prominent concentrate is required on issues of pertinence to IPR and on the portrayal of microbial strains required in customary fermentation processes. Accentuation must be put on IPR training for researchers. National governments should set up the imperative foundation for IPR to encourage the procedure. At the institutional level, this framework would incorporate innovation administration workplaces for helping researchers in methods identifying with protected innovation matters. The procedures utilized as a part of the more propelled regions of farming biotechnology are by and large secured by IPR, and the rights are by and large claimed by parties in created nations.

Communication and consumer perceptions

Correspondence between different partners is basic in proactively captivating with purchasers. Correspondence must be set up with the general population everywhere on prepared food and related risks. Correspondence step by step manufactures certainty and will be basic to propelling the utilization of biotechnologies in food preparing and security. The essential part of correspondence in this regard is to guarantee that data and conclusions from all partners are fused in the exchange and basic leadership process. The requirement for particular gauges or related writings and the strategies took after to decide them ought to likewise be obviously sketched out. The procedure, in this manner, ought to be straightforward. Open mindfulness and instruction are basic to the accomplishment of food bioprocessing and food wellbeing in developing nations. More prominent consideration must be coordinated toward understanding shopper and maker (processor) recognitions on food wellbeing and quality in developing nations.

On the off chance that foods are to be advanced as being sheltered and sound, their dietary and wellbeing qualities must be straightforwardly shown by displaying logical information to substantiate the nourishing and medical advantages and by applying great assembling/cleanliness practice and HACCP as security measures to guarantee that issues of buyer concern are tended to.

Technical capacities and technology transfer

Conventional fermented foods ought to be seen as profitable resources. Governments ought to profit by these benefits and increase the value of them by supporting examination, training and advancement, while expanding on and building up the indigenous learning base on food fermentations. Government offices in creating nations should concentrate on the improvement of specialized abilities to manage developing specialized issues.

The specialized limits of scholastic and research organizations ought to be fortified in the fields of food biotechnology, food preparing, bioprocess building and food wellbeing through preparing and trade programs for analysts. Such projects ought to underscore coordinated effort with both developed and developing nation foundations occupied with chip away at food biotechnology, starter culture advancement, bioprocess building and food security.

Preparing capacities in food biotechnology and food wellbeing ought to be produced inside developing nation establishments through the presentation of degree courses keeping in mind the end goal to widen the in-nation specialized help base for food bioprocess improvement. Given the likenesses among fermentation procedures crosswise over locales, a stock of organizations occupied with food biotechnology in developing nations would be an advantage in encouraging systems administration among foundations. food processors, policymakers and hardware makers ought to likewise be coordinated into the systems administration exercises.

The advancement of suitable levels of bioreactor innovation with control bioprocess parameters will be important to enhance the clean states of the fermentation procedures. Research and infrastructural improvement to empower the practical generation of characterized starter societies in a steady configuration (i.e. societies which don't require refrigeration and have drawn out time span of usability under encompassing conditions) ought to be organized.

Framework advancement to encourage the exchange and adjustment of fermentation innovations created at the research facility level to the family unit and town and, where essential, the undertaking level ought to be organized. Suitable levels of hardware will likewise be required to encourage the downstream handling of these items. Traceability frameworks that encourage the separation and distinguishing proof of food items ought to be organized so as to expand advertise open doors for these items. A natural pecking order way to deal with guaranteeing food wellbeing ought to be organized by governments.

Food security administration frameworks ought to be fortified by executing orderly food wellbeing measures, for example, GHP, GMP and HACCP in food fermentation operations. Symptomatic packs are imperative instruments for observing and checking the level of sanitation in preparing plants.

Exceedingly delicate and fast symptomatic packs are priceless for observing and quickly identifying concoction and microbiological perils that represent a risk to human wellbeing, with high accuracy and affectability. The improvement of minimal effort symptomatic packs reasonable for use by little processors would significantly encourage food security observing. Advancement should focus on the acknowledgment of multiplex demonstrative frameworks with the ability to recognize a few pathogens or numerous substance contaminants utilizing a solitary analytic pack. The improvement of demonstrative units at a national level could additionally decrease their cost of generation. Given the local specificity of bacterial pathogens at the species and subspecies levels, such analytic packs ought to be created with specificity and affectability to the species or subspecies that are common in a particular district. Speculation is in this way required for the improvement of aptitude, offices, and the framework for the large-scale manufacturing of antibodies, cell culture innovation and for the development of specialized know-how on gathering the essential parts of demonstrative units.

The advancement of national danger - profile databases that report the common pathogens in various districts will be basic. Such data would be helpful for additionally look into the advancement of indicative units with high exactness and affectability and in executing HACCP and hazard evaluation inquire about. The way of life accumulation of recognized irresistible specialists in the risk profiles could assume a vital part for particular counter acting agent generation for use in the improvement of immunoassay analytic packs.

Recognizing needs for activity for the worldwide group

Market strengths have been the significant drivers of progress in the food segment of developing nations. Current biotechnological apparatuses are probably going to assume a more noteworthy part in the improvement of productive science-based procedures for food preparing and wellbeing keeping in mind the end goal to react to shopper request. The generation of high-esteem aged items, for example, chemicals, useful food fixings and food added substances is probably going to keep on increasing in creating nations. The universal group (FAO, UN associations, NGOs, giversgivers, and improvement offices) can assume a noteworthy part in helping creating nations to expand the advantage to be gotten from food bioprocessing. The reception of biotechnology-based strategies in food preparing and for food wellbeing and quality checking is subject to a few factors that incorporate limit working in specialized and administrative zones, arrangement definition, administrative structures and territorial systems. On the premise of investigation in this archive, various need regions to be bolstered by the worldwide group are introduced underneath:

Capacity building and human asset advancement

- Support for fundamental and propelled education.
- Prioritization of particular regions for speculation.
- Advancement of strategies, needs and activity programs that advance food fermentation as a method for tending to food security.
- Support for human asset advancement in a scope of logical controls – food biotechnology, food security, bioengineering and chemical innovation.
- Support for limit building activities for family unit level, little and medium-scale processors of fermented food.
- Support for IPR improvement.

Innovation exchange and support for innovative work

- Change of the significance of national research to the requirements of the food division in creating nations.
- Improvement of intensity and the formation of an empowering domain that is helpful for private-part interest in research, advancement and development for the overhauling of food maturation procedures to react to showcase request.
- Foundation and fortifying of the examination and infrastructural bolster construct for work with respect to starter culture improvement, bioreactor plan and for the advancement of indicative gear for checking food security and traceability. This infrastructural bolster base would incorporate labs, lab gear, cell bank offices for the best possible protection and capacity of microbial culture arrangements.
- Advancement of logical information to substantiate the nutritious, wellbeing and medical advantage claims related with fermented food.
- Foundation of pilot preparing offices for the scale-up and testing of advancements created so as to encourage their appropriation.

Systems administration and groups

- Support for the advancement of administrative systems for food safety.
- Support for North-South and South-South preparing and trade on food biotechnologies, bioprocess engineering and food safety.
- Advancement and assistance of systems administration among researchers, analysts, little and medium scale food processors and the retail part to encourage learning and data sharing.
- Support for utilizing the conventional information base in the updating of food fermentation preparing operations.

REFERENCES

1. Bronzwaer, S. (2008). Trends in Food Science & Technology. [pdf] Available at: http://www.e-ijaet.org/media/21I12-IJAET0112112-FOOD-PROCESSING.pdf [Accessed 29 Jul. 2017].
2. Mathavi, V. (2013). New Trends in Food Processing. [pdf] Available at: http://www.e-ijaet.org/media/21I12-IJAET0112112-FOOD-PROCESSING.pdf [Accessed 29 Jul. 2017].

INDEX